CONTEMPORARY CASE

Natural Hazards & Disasters

David Holmes and Sue Warn

Philip Allan Updates, an imprint of Hodder Education, an Hachette UK company, Market Place, Deddington, Oxfordshire OX15 0SE

Orders
Bookpoint Ltd, 130 Milton Park, Abingdon, Oxfordshire, OX14 4SB
tel: 01235 827827
fax: 01235 400401
e-mail: education@bookpoint.co.uk

Lines are open 9.00 a.m.–5.00 p.m., Monday to Saturday, with a 24-hour message answering service. You can also order through the Philip Allan Updates website: www.philipallan.co.uk

ISBN 978-1-84489-612-7

First printed 2008
Impression number 8 7 6 5 4
Year 2014 2013 2012 2011

Front cover photograph reproduced by permission of NOAA

Printed in Italy

Hachette UK's policy is to use papers that are natural, renewable and recyclable products and made from wood grown in sustainable forests. The logging and manufacturing processes are expected to conform to the environmental regulations of the country of origin.

Contents

Part 4: Geomorphic hazards

Part 5: Biohazards

Part 6: Multiple hazard zones

Part 7: Responding to hazards and disasters

Part 8: Global trends in the occurrence and impact of hazards

Part 9: Advice to students

Introduction

Few so-called natural **hazards** are entirely natural, and their realisation into **disasters** is the result of human **vulnerability**. Some hazards, such as wildfires (see pp. 61–66), are naturally occurring **biohazards**, or they can be caused by the direct and indirect impact of human actions. These are therefore classified as **quasi-natural** hazards. In many cases, human actions intensify the impact of natural hazards — for example, exacerbating flood risk. Some apparently natural events have an entirely human cause — for example, floods resulting from dam destruction, or the Indonesian mud volcano, which resulted from gas drilling (see p. 17).

The hazard profiles (Figure 1) show the relative dimensions of the hazard according to six indicators:

- **magnitude** — the size of the event; for example, wind speed on the Beaufort scale, maximum height or discharge of a flood, or the size of an earthquake on the Richter scale.
- **speed of onset** — this is similar to the 'time lag' on a flood hydrograph. It is the time difference between the start of the event and the peak, and ranges from rapid events (e.g. the Kobe earthquake) to slow events (e.g. the drought in the Sahel of Africa).
- **duration** — the length of time that the environmental hazard exists. This varies from a matter of hours (tornado) to decades (drought).
- **areal extent** — the size of the area covered by the hazard. It can range from small scale (an avalanche chute) to continental (drought).
- **spatial predictability** — the distribution of the hazard; whether it occurs in particular locations (e.g. plate boundaries) or is widely dispersed across the world.
- **frequency** — how often an event of a certain size occurs. For example, a flood of height 1m may occur, on average, every year on a particular river; a flood of height 2m might occur only every 10 years. The frequency is sometimes called the recurrence interval.

About this book

Part 1 begins by defining hazards and disasters and distinguishing between them in the contexts of risk and vulnerability. Parts 2–5 contain case studies of particular hazard types. The general characteristics for each hazard type are profiled

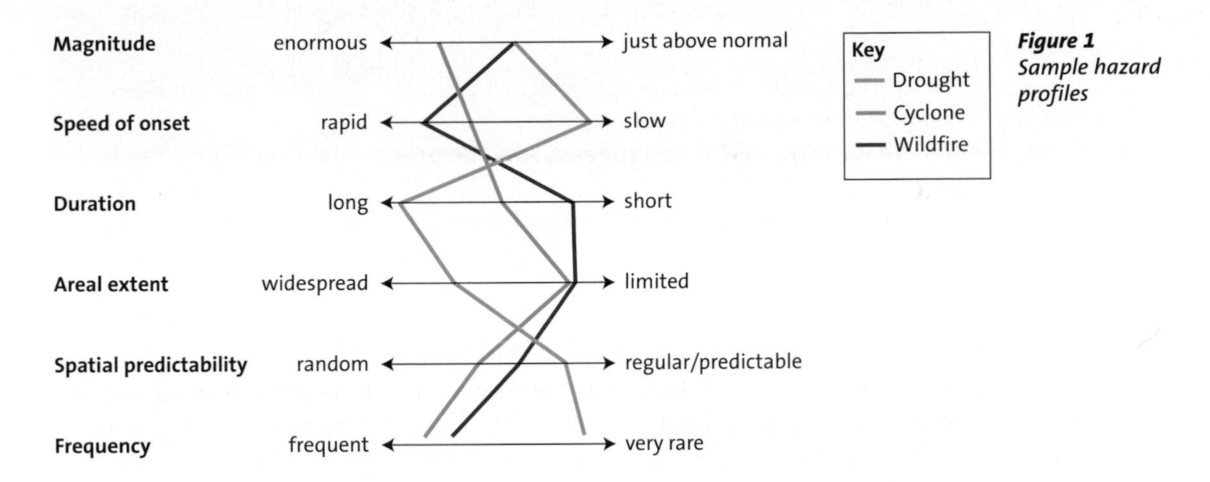

Figure 1
Sample hazard profiles

as shown in Figure 1. The causes are summarised and the spatial distribution analysed. Case studies of contrasting hazard and disaster events, including recent examples, are featured. Understanding of the key aspects of each hazard type is tested by inviting an assessment of the relative impacts of particular physical or human factors, or an evaluation of the success of management strategies employed before, during and after the event. Strategies are classified as:

- modify the event
- modify vulnerability
- modify the loss

In Part 6, the issue of multiple hazard hotspots is explored. This is followed in Part 7 by an overview of the latest techniques used to manage hazards and disasters, ranging from **community preparedness** and education to the **techno-fix**. The impact of the state of development on management strategies and the roles of key players are evaluated, incorporating many of the case studies used in the preceding chapters.

Part 8 is an overview of **global trends**. It assesses the evidence as to whether the world is becoming a more hazardous place, and whether disasters are becoming more numerous, by analysing trends in the number of events and their social and economic impacts. Links to the **context** hazard of global warming and its relationship to both hazard magnitude and frequency are also evaluated.

Part 9 provides guidance on a number of problems that students experience when answering examination questions on hazards. These include:

- researching hazards effectively using books, articles and websites
- developing effective techniques for organising notes
- managing the wealth of data when studying a mega event such as the 2004 Boxing Day tsunami
- drawing effective maps and diagrams under the time pressures of exams, for example to show the impact of types of plate boundary
- planning and writing quality hazard essays

Question

(a) Write a brief description of the main characteristics of cyclones, droughts and wildfires, according to their profiles shown in Figure 1.
(b) Read the section on wildfires (p. 61) and suggest why the profile shown in Figure 1 is very generalised.
(c) Assess the strengths and weaknesses of hazard profiles for comparing (i) hazard types and (ii) hazard events.

Guidance

(a) Study the profiles and refer to the relevant chapters for an explanation of any differences.
(b) Wildfires come in a huge variety of sizes and have many causes. Take the features of the generalised profile and develop the point.
(c) The hazard profile is a qualitative tool, but it is useful for profiling different types of hazard where there is a clear pattern. Equally, profiles can be used for events where the magnitude is measurable. For an assessment, explain the problems with its use.

Key terms

Asthenosphere: a layer in the Earth's mantle that lies below the **lithosphere**. It has a higher viscosity and is more resistant to deformation than the lithosphere

Benioff zone: a sloping plane formed where oceanic plates are **subducted** beneath an overriding continental plate. It may be the site of intermediate/deep-focused earthquakes.

Desertification: the spread of desert-like conditions to non-desert areas.

Drought: an extended period of time with exceptionally low precipitation.

Earthquake: a series of vibrations and shockwaves initiated by volcanic eruptions or movements along plate boundaries, which can occur at **destructive**, **constructive**, or **conservative** margins.

Epicentre: the point in the Earth's surface directly above the origin or focus of an earthquake.

Flood: occurs when a river's **discharge** exceeds the capacity of the channel to carry it (i.e. **over bankfull**). The **recurrence interval** is the average interval between occurrences of two events of similar scale/magnitude plotted on a **flood-frequency curve**.

Hurricane: a tropical storm that occurs in the Atlantic or Caribbean oceans; a **typhoon** is a tropical storm that occurs in the Indian or Pacific oceans. Both are **cyclones**, i.e. low-pressure systems accompanied by severe weather. The magnitude and intensity of a hurricane is measured on the **Saffir–Simpson** scale.

Lahar: a mudflow, as heavy rain mobilises ash from eruptions.

Lava: magna **extruded** onto the surface. The variations in silica content lead to acid, intermediate and basic lava. Volcanoes can be **active** (erupted within recorded history), **dormant** (no sign of recent activity but not deeply eroded), or **extinct** (no recorded eruptions, deeply eroded). Volcanoes may be basic shield (basaltic lava), composite (lava and ash), acid dome (viscous acidic lava), ash/cinder cone (pyroclastic material) or **caldera** — a large **crater** formed from a cataclysmic eruption.

Liquefaction: the rapid fluidisation of sediment as a result of an earthquake.

Magma: the molten material under the surface of the Earth, which has risen from the **mantle**.

Richter scale: a logarithmic scale measuring the magnitude of an earthquake. The severity of ground shaking or **intensity** is measured by the **Mercalli** scale.

Risk assessment: the probability that a hazardous event of a particular magnitude will occur within a given period, and thus estimating its impact and the community's **vulnerability**.

Seismograph: an instrument that measures earthquake movements, recording **primary** (p) fastest waves, **secondary** (s) waves and **surface** (l) waves, which are the slowest.

Storm surge: an abnormal, temporary rise in sea level due to air being sucked up by low pressures.

Tornado: a violent atmospheric storm, usually small in size and short-lived but capable of intense damage. Tornadoes are measured on the **Fujita** intensity scale or the TORRO scale in the UK.

Tsunami: a long-wavelength ocean wave, generated by a sudden displacement of the sea floor. It can lead to a retreat of the sea level as a tsunami wave trough approaches the shore, known as **draw down**. The **run-up stage** is when a tsunami makes landfall.

Volcanic plume: results from the upwelling of magma from a **hot spot** within the mantle.

Websites

There are a number of good websites on hazards. The following list shows some of the best ones.

- **www.gesource.ac.uk/hazards/** is a high-quality, general natural hazards site. Check out the excellent maps, e.g. 'billion dollar' disaster events and world climate maps.
- The Benfield Hazard Research Centre at **www.benfieldhrc.org/index.htm** has some useful downloadable publications, e.g. *Hazard and Risk Science Review*, and a good newsletter.
- **http://quake.wr.usgs.gov/** is the USGS website on all matters related to earthquakes. It is a good starting point for research. Look out for general earthquake information — you can download some good material.
- The European Strong-Motion Database is a rich source of information. Find it at **www.isesd.cv.ic.ac.uk**. (You have to register but the data are good.)

- **http://volcano.und.edu/** takes you to the Volcano World homepage. There is a feast of information for you to check out, including up-to-date eruptions.
- **www.ess.washington.edu/tsunami/index.html** is a general tsunami site hosted by the University of Washington. Find the link to the animations.
- **www.nhc.noaa.gov** is the website of the US Hurricane Center. It is probably the best general climate site for hurricanes and tornadoes. It has current watches and warnings.
- The Dartmouth Flood Observatory at **www.dartmouth.edu/~floods/** has maps and tables — both historic and up-to-date — for all major global flood events.
- **www.ceh.ac.uk/data/nrfa/index.html** is the website of the National Rivers Flow Archive. You can download detailed daily river data for UK rivers.
- The USGS website has a selection of hazards fact sheets. See the link at **http://water.usgs.gov/wid/index-hazards.html**.
- The British Geological Survey has some good information. Find it at **www.bgs.ac.uk/**. The link Britain Beneath Our Feet has some good hazard maps for the UK.
- Swiss Re (**www.swissre.com**) and Munich Re (**www.munichre.com**) are global insurance companies. Check out their publications sections for technical documents relating to a range of hazards.
- The Natural Hazards Center at **www.colorado.edu/hazards/** has some useful publications that can be downloaded.
- The International Red Cross homepage is at **www.ifrc.org/**. World disaster reports can be obtained here.

Part 1

Dimensions of disaster

A **hazard** can be defined as 'a perceived natural/geophysical event that has the potential to threaten both life and property' (Whittow). The geophysical event would not be hazardous without, for example, human occupancy of its location. That is to say, river or coastal **floods** would not be hazards if people did not live in river floodplains and coastal plains. Since hazards occur at the interface between natural and human systems, it is unlikely that any hazard is truly natural. There is a continuum from natural to quasi-natural (na-tech) to man-made (Figure 2).

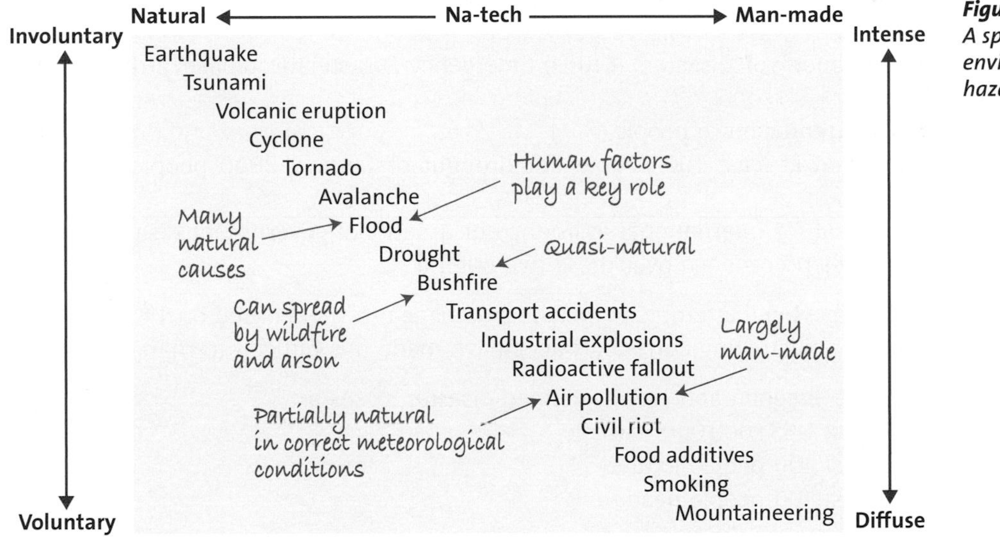

Figure 2
A spectrum of environmental hazards

There is mounting evidence that worldwide environmental changes, especially those associated with the enhanced greenhouse effect, will produce large-area hazards (**context** hazards) as opposed to **site-specific** hazards, because they will exacerbate atmospheric hazards, such as storms and floods, and facilitate the spread of diseases such as malaria.

A **disaster** is the realisation of a hazard, when it 'causes a significant impact on a vulnerable population' (Dregg, Figure 3). The terms hazard and disaster are often used casually or synonymously.

The main issue is to establish a threshold for 'significant impact' in terms of numbers killed or affected. For inclusion in the Centre for Research on the

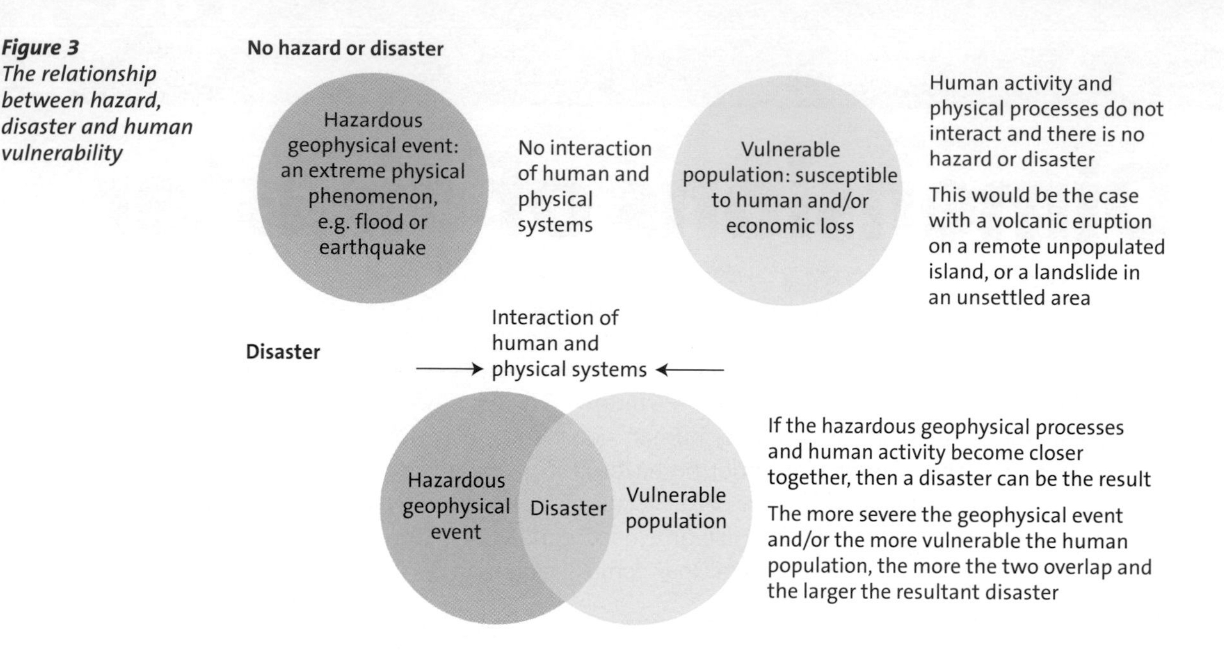

Figure 3
The relationship between hazard, disaster and human vulnerability

Epidemiology of Disasters (CRED) Emergency Disasters Database, a disaster must have:

- killed ten or more people
- affected at least 100 people (for **drought** or famine, 2000 people have to be affected)

An appeal for international assistance or a national government disaster declaration takes precedence over these two criteria.

There have been attempts to include a damage threshold based on 1% of GDP lost. However, fluctuating currency values have made this difficult to manage.

In 1969, Sheenan and Hewitt defined disaster losses as:

- at least US$1 000 000 damage
- at least 100 people injured
- at least 100 people dead

For the event to be defined as a disaster, one or more of these criteria had to be satisfied.

In 1990, the reinsurance company Swiss Re defined disaster losses as either or both of:

- at least 20 people killed
- insured damage of at least US$16.2 million

Risk is defined as 'the probability of a hazard occurring and creating loss of lives and livelihoods'. It might be assumed that risk to hazard exposure is involuntary, but in reality people may consciously place themselves at risk. Reasons for this include the following:

- Hazards are unpredictable. It is difficult to know when or where an event will occur and what the magnitude of the event will be.
- Changing risks: natural hazards vary in space as well as through time because

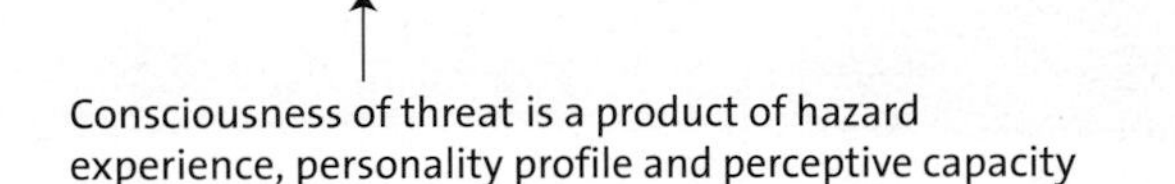

Figure 4 *The risk perception process*

of changing physical factors and human activities. For example, a semi-extinct volcano such as Mount St Helens was not expected to cause a major eruption. With the rise of sea level, places that were once safe are now at risk — for example, low-lying coastal plains are more prone to storm surges and floods. Deforestation of watersheds could lead to less interception and more flashy hydrographs, thus increasing the frequency and magnitude of flood events.

- People stay in hazardous locations because of a lack of alternatives. The most vulnerable and poverty stricken people are often forced to live in unsafe locations, such as hillsides, floodplains and regions subject to drought. There may be economic reasons linked to their livelihoods, such as subsistence farming. A shortage of land or a lack of knowledge of alternatives may promote stability as it is never easy to uproot and 'risk' a move to another location.
- Many people subconsciously weigh up the benefits versus the costs. Fertile farming land on the flanks of a basaltic volcano, or on alluvium-covered floodplains, or the attractive Californian climate outweigh the risks from floods, eruptions or earthquakes.
- Perceptions of hazard risks tend to be optimistic. Individuals accept that hazards are part of everyday life or result from 'acts of God'. They also have confidence in the technological fix. They seek solace in statistics, such as those published in the USA, which show that each year only 1.8% of US households are affected by **floods**, **hurricanes**, **earthquakes** or **tornadoes**/windstorms, and that the risk of death is far lower than that from influenza or car accidents. People perceive regularity in irregular events. For example, they may speculate that once a hazard event has occurred, they will be safe for the next few years. Most people consider themselves safe from any event with a recurrence level of more than 100 years.

Figure 5 *Relationship between severity of hazard, its probability and the degree of risk*

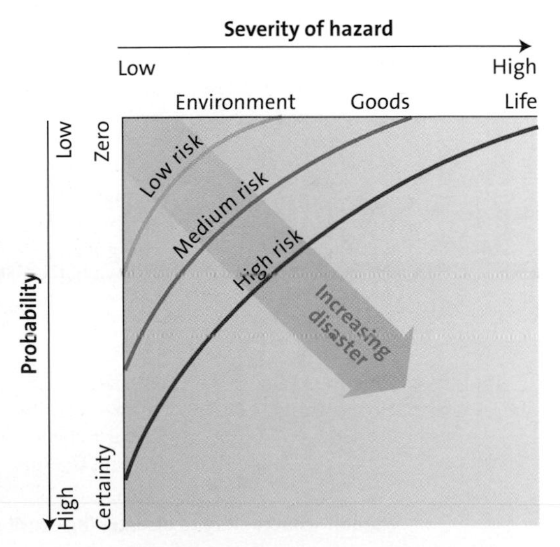

Risk assessment defines the likelihood of harm and damage. Figure 5 shows the theoretical relationships

Figure 6
A fishing village in Bangladesh — poorer people are among the most vulnerable to hazard events

between the severity of a particular environmental hazard, the probability of occurrence and the degree of risk. The degree of risk varies with the type of hazard and its profile of predictability, but risk is also altered by human conditions and actions. For instance, two earthquakes of similar magnitude on the **Richter scale** (Bam and California) can have very different consequences because the lives of people in poorer, less developed countries are generally at greater risk than those in MEDCs.

Vulnerability implies 'a high risk of exposure to hazards combined with an inability to cope'. In human terms it is the degree of resistance offered by a social system to the impact of a hazardous event. This depends on the resilience of individuals and communities, and the reliability of management systems that have been put in place.

Conditions of poverty and low economic status amplify vulnerability. The young, the old, especially women, and the chronically malnourished are more likely to be killed by hazards.

Figure 7
Risk–vulnerability relationship

Figure 7 explores the relationship between risk and vulnerability. LDCs suffer from high risk combined with low security (i.e. high vulnerability).

Figure 8 summarises how vulnerability occurs. Although it is a feature of the world's poorest countries, it is significant in LEDCs that are experiencing rapid demographic and economic change to lower middle income status. There are also significant concentrations of highly vulnerable people in MEDCs, especially in poor urban areas (see *Case study 26*, p. 94).

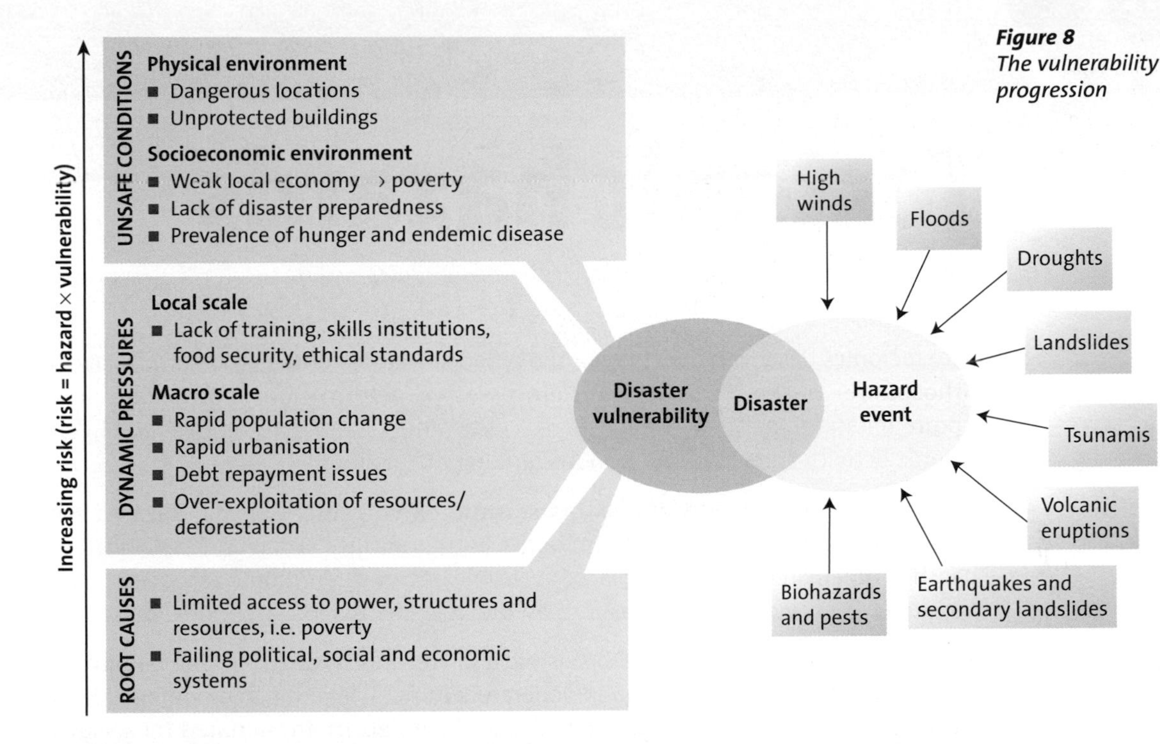

Figure 8 *The vulnerability progression*

2 Using case studies

Question

Study Figure 9.

Hazards increasing		Increasing vulnerability		Decreasing capacity		
■ The unsustainable use of fossil fuels is warming the planet ■ The resulting change in climate is increasing the frequency and severity of weather- related hazards (e.g. floods, droughts, wind-storms) and expanding the range of disease vectors	×	■ Hazards only become disasters when people get in the way (exposure) ■ Unsustainable development involves poor land use (e.g. building on floodplains, unstable slopes, lowland coasts) and environmental degradation (e.g. bleaching of coral reefs, destruction of coastal mangroves, deforestation of water catchments), which are increasing vulnerability by putting millions more in harm's way	÷	■ Vulnerable communities need skills, tools and money to cope with the effects of climate change ■ Debt repayments, inequitable trade arrangements, selective foreign investment and the redirection of aid funds towards geostrategic regions mean that the poorest and most vulnerable communities lack the resources to cope ■ Migration of millions from rural to urban areas in the hope of finding work and avoiding disaster is undermining traditional coping strategies	=	**Increasing risk** for many of the world's people, especially in LDCs

Figure 9 *The risk equation*

(a) Explain, with examples, how unsustainable development increases the risk equation.

(b) Suggest why the risk equation will change in the future, leading to an increasing risk of disaster.

Guidance

(a) Look at hazards, vulnerability and capacity as shown in Figure 9.

(b) Consider physical factors (e.g. climate change) and human and economic factors (failure to 'make poverty history', increasing LEDC mega-city urbanisation).

Part 2

Tectonic hazards

Plate tectonics describes the study of the broad structures of the rigid surface layers (**lithosphere**) of the Earth and the processes of deformation (e.g. folding) and rupture (faulting) that give rise to them. A clear understanding of the geography of tectonic activity is central to an understanding of this topic.

Tectonic hazards can be classified as **seismic** or **volcanic**. Seismic hazards are generated when rocks within 700 km of the Earth's surface come under such stress that they break and become displaced. Volcanic hazards are mainly associated with eruptions. Their violence is measured by the volcanic explosivity index (VEI).

Secondary hazards associated with seismic and volcanic hazards include **tsunamis**, debris flows (**lahars**) and eruption-generated rainfall leading to flooding. Many geomorphological hazards, such as landslides (see Part 4) are initiated by seismic events, for example ground shake or volcanic disturbance.

Earthquakes

The crust of the Earth is mobile, so there is a slow build-up of stress within rocks. When the pressure is suddenly released, parts of the surface experience an intense shaking motion that typically lasts for only a few seconds. This is the earthquake. The plane of rupture is called a fault and the location of movement is the hypocentre or focus. The point immediately above the focus on the land surface is called the **epicentre**. The depth of focus is important in determining the amount of surface damage that results. Three broad categories are recognised:

- deep focus, 300–700 km
- intermediate focus, 70–300 km
- shallow focus, 0–70 km

Shallow-focus earthquakes cause the greatest amount of damage and account for approximately 75% of all earthquakes.

The majority of earthquakes occur along plate boundaries (**interplate** — see Figure 10). **Intraplate** earthquakes occur less frequently, within plate interiors. The most powerful earthquakes are found at destructive, interplate margins.

The distribution of earthquakes reveals the following pattern of tectonic activity:

- the **oceanic fracture zone** (OFZ) — a belt of activity through the oceans along the mid-ocean ridges, coming ashore in Africa, the Red Sea, the Dead Sea rift and California

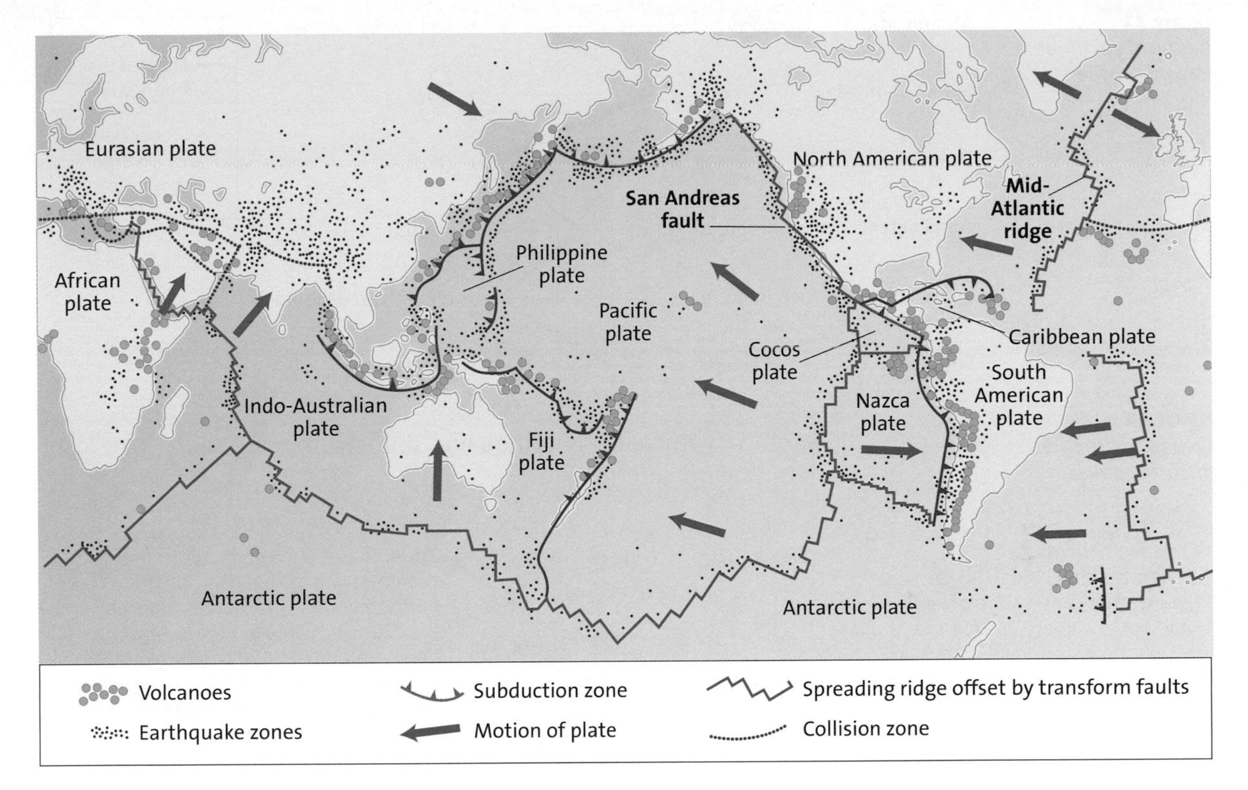

Figure 10 *Global distribution of earthquakes and volcanoes*

- the **continental fracture zone** (CFZ) — a belt of activity following the mountain ranges from Spain, via the Alps, to the Middle East, the Himalayas to the East Indies and then circumscribing the Pacific
- scattered earthquakes in continental interiors (associated with volcanic islands)

There are three types of plate boundary:

- **conservative** (oblique-slip or sliding) margins, where one plate slides against another — this type of boundary is rare but well developed in the San Andreas fault in California
- **divergent/constructive** margins, most clearly displayed in the mid-ocean ridge systems of the OFZ
- **destructive** (convergent or subductive) margins are well developed along the CFZ

Destructive boundaries cause the most earthquakes.

The initial impact of an earthquake is ground shaking. The severity of this depends on the magnitude of the earthquake, the distance from the epicentre and the local geological conditions. At a local scale, the intensity of ground shaking can be increased by the surface topography and nature of the surface materials. For instance, hills and escarpments focus earthquake waves, whereas surface sands, gravels and saturated materials amplify earthquake waves.

The impact is also determined by human factors, such as population density, the level of development, community preparedness, design of buildings and level of prediction. Earthquake impact is greatest in cities where building collapse leads to the loss of human lives.

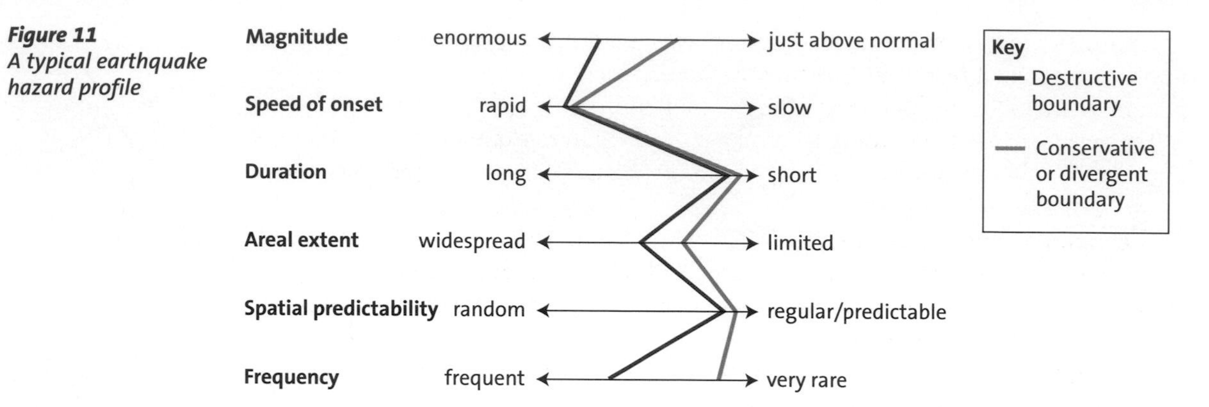

Figure 11
A typical earthquake hazard profile

Figure 12
Secondary impacts of earthquakes

It is often the secondary impacts of earthquakes that result in the most significant losses of life and infrastructure (Figure 12).

Mass movements (see p. 54)
Avalanches and landslides are slope failures that can result from ground shaking. In January 2001, an earthquake measuring 7.6 on the Richter scale occurred near El Salvador. It hit the suburban town of Las Colinas and triggered a landslide, which killed over 500 people. This highlights the problem of increasing urbanisation, especially in areas of high seismicity — see p. 58.

Tsunami (see p. 20)
Giant sea waves caused by offshore earthquakes may be devastating (see p. 22). During the twentieth century, approximately 50 000 people living in countries of the Pacific Rim lost their lives to tsunamis. This figure does not include the 2004 event.

Soil liquefaction
Soils with a high water content lose their mechanical strength when violently shaken and start to behave as a fluid. The 1985 earth- quake in Mexico City produced seismic waves that devastated the city, which was built on ancient lake sediments.

Fire
In urban areas, one of the most destructive consequences of a major earthquake is fire. The Great Kanto earthquake that struck Tokyo in 1923 overturned tens of thousands of small stoves that soon merged into a great firestorm — 200 000 people died and 360 000 buildings were destroyed.

General effects on people and the environment
The range of impacts includes: collapse of buildings; destruction of road systems, railways, bridges and other forms of communication; destruction of services provision such as gas, water and electricity; flooding; food short- ages; disease; disruption of the local economy (either subsistence or commercial). Some impacts are short term, others have an effect over a longer period.

THE EARTHQUAKE IN KASHMIR, 2005

On 8 October 2005, an earthquake measuring 7.6 on the Richter scale devastated parts of northern Pakistan and neighbouring Kashmir.

Although the region is geologically active, it had suffered relatively few large earthquakes in the last century. This meant it was 'primed' for a big one. The affected region lies along a fault line where the northward-moving Indian tectonic plate plunges beneath the Eurasian plate, pushing up the Himalayan mountain range. GPS measurements have recorded the plates moving towards each other at a rate of about 2 cm per year.

Pakistan's prime minister, Shaukat Aziz, described the event as 'a disaster of unprecedented proportions in Pakistan's history'. Around 55 000 people died and 3.5 million lost their homes.

A number of physical, human and geo-political factors conspired to create what was dubbed by some observers as 'the mountain tsunami':

- Unseasonable torrential rain hampered relief efforts.
- The event occurred in October when the region began to experience a marked drop in temperature. Priorities had to be shifted from rescuing the injured to providing shelter for the survivors.
- The mountainous Himalayan terrain caused problems. The area lacked a good road infrastructure. Helicopters could not land easily in the remote locations.
- India and Pakistan are 'young' countries — many of the dead were children, and many children were left orphaned.
- Many of the people affected by the earthquake were poor and unprepared. They had no resources to fall back on.
- One reason why so many people died was because the schools and hospitals were shoddily built and collapsed easily.
- The Pakistani government was criticised for its slow response to the disaster.
- There has been criticism of the coordination between military and civilian authorities. However, coordination was complicated due to the large number of organisations and individuals involved.
- India offered aid but Pakistan declined the loan of helicopters to assist the rescue effort because the earthquake was located in a highly sensitive disputed area.

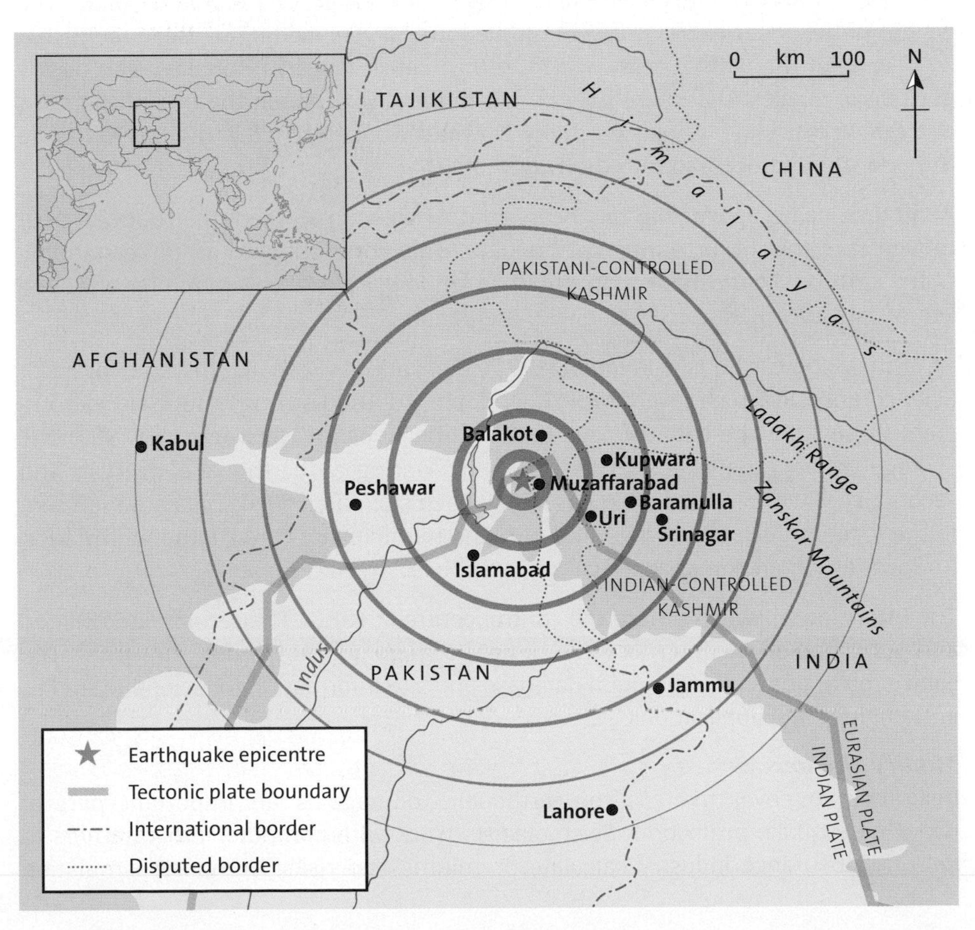

Figure 13
The earthquake in Kashmir, 2005

Managing the earthquake hazard

Modify the event

During the 1970s and 1980s there was a general feeling that the technological capability and engineering skills to control earthquakes would soon be developed, for example by lubricating the fault plates. Unfortunately this has not proved possible. Seismic activity cannot be controlled and probably never will be.

Modify vulnerability

Vulnerability can be reduced by earthquake-proofing buildings and infrastructure. New buildings and structures can be built to resist shaking (**aseismic**). It is expensive and costs escalate with increasing resistance. Making existing buildings earthquake proof (retrofitting) is even more costly.

Scientists know the areas where earthquakes can occur. The characteristic magnitude–frequency distributions are also known, so it is possible to predict earthquake occurrence. However, the inability to forecast earthquakes *exactly* limits the potential warnings and evacuation procedures.

Some scientists believe they may be able to predict earthquakes in the longer term. For example, one theory is that on the San Andreas fault, rain falling early in the year seeps through pores in the upper layers of rock, raising the water table. The weight of the water exerts pressure on the lower layers, adding stress to faults that lie deeper in the Earth's crust, which in turn can trigger earthquakes. The theory goes on to suggest that there may be a 5 month lag between the end of the rains and the increase in earthquake activity, as water would take a few months to diffuse through the bedrock and into the water table.

Remote sensing by satellite has been used to view small (centimetre level) crust movements that occur during earthquakes. This contributes to an understanding of the earthquake mechanism itself but so far, low levels of resolution have limited use of this information in earthquake prediction.

After the Kobe earthquake in Japan (1995), the National Research Institute for Earth Science and Disaster Prevention (NIED) deployed 1000 strong-motion accelerometers throughout the country. This is called the **Kyoshin Network**, or K-NET. The average distance between stations is 25 km. During an earthquake, primary and secondary wave velocities are measured at each site and logged. Data are then sent to the local municipality (via a modem). The municipality can use the information for local emergency management and response.

The information is also sent to the control centre at NIED. The records are collated and published on the internet (**www.k-net.bosai.go.jp/k-net/index_en.shtml**). The centre maintains a strong-motion database and site information for scientific studies and engineering applications.

Modify the loss

Insurance to cover the cost of earthquake damage is an important part of wider earthquake protection. Seismologists work with computer programmers to help the insurance industry calculate premiums and risks. Computer simulation technologies are used to estimate the probability of damage from different earthquake events. There are three types of information:

- seismicity — the raw information about how frequently earthquakes affect a location
- seismic hazard — the probability that a certain strength of shaking will occur
- seismic risk — the probability that a certain amount of risk will occur (see Figure 14)

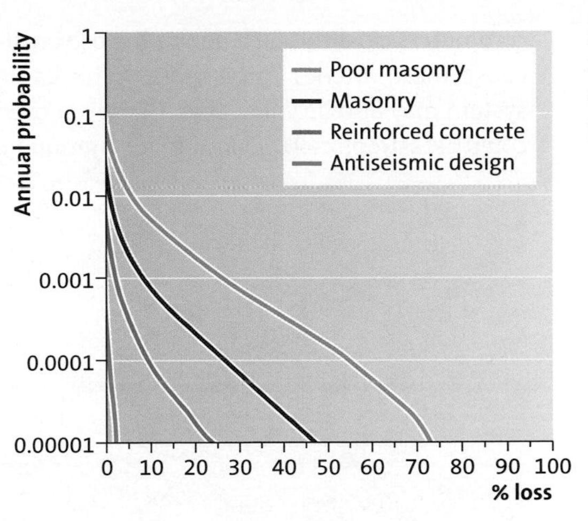

***Figure 14** Seismic risk curves by building type for a town in western Turkey*

IS SAN FRANCISCO READY FOR THE NEXT 'BIG ONE'?

Case study 2

Just before dawn on 18 April 1906, the ground shook for over 1 minute. San Francisco was home to 4 000 000 people, 25% of the population of the USA west of the Rocky Mountains. The earthquake measured 7.9 on the Richter scale. It caused 300 deaths and 225 000 were made homeless. This earthquake and the secondary firestorm produced one of the worst urban disasters of the twentieth century. It was initiated by a complete rupture of the 500 km northern section of the San Andreas fault, tearing apart at an estimated 13 000 km h^{-1}.

In the relative seismic silence since 1906, the city has been rebuilt and heavily urbanised. The San Francisco Bay area plays an important part in the US economy. Over the past 100 years, tectonic activity in the area has been storing strain, making the next big earthquake a question of when, not if (Figure 15).

Yet many scientists and planners claim that the city is poorly prepared. It is feared that, without better preparation, an earthquake of similar magnitude to the 1906 event would unleash horrors to rival those of a century ago. In particular:

- Nearly 40% of the city's private buildings could be destroyed, since a vast number of them pre-date modern building codes. More than half the city's housing sits above a 'soft storey' — that is, a lower floor comprising a garage or shop with large openings that offer little structural support during shaking.
- As the city has expanded, many structures are built on reclaimed land from the bay area, which is particularly prone to **liquefaction**.
- The city's lowest-income residents are most at risk. More than 60% of San Franciscans rent, but landlords have little incentive to make their buildings safer as they cannot pass on the costs of retrofitting to these tenants.
- Fire remains a considerable danger, particularly in areas where gas is supplied to parts of the city that are dense with wooden housing. There is also concern over the integrity of water mains to put out fires. In the Loma Prieta earthquake in 1989 (100 km south of San Francisco) water pipes were ruptured and a massive holding tank was drained in minutes.
- The transport infrastructure is another concern. Since 1906, the population of San Francisco has doubled. Suburbs have developed over many Bay Area faults and

commuters cross this area during the peak rush hour. A repeat of the 1906 event would cause these arteries to be broken. The Bay Area Rapid Transit (BART) underground system may also be vulnerable. The soil above the subway is prone to liquefaction and could catastrophically flood the underground in a major shake.

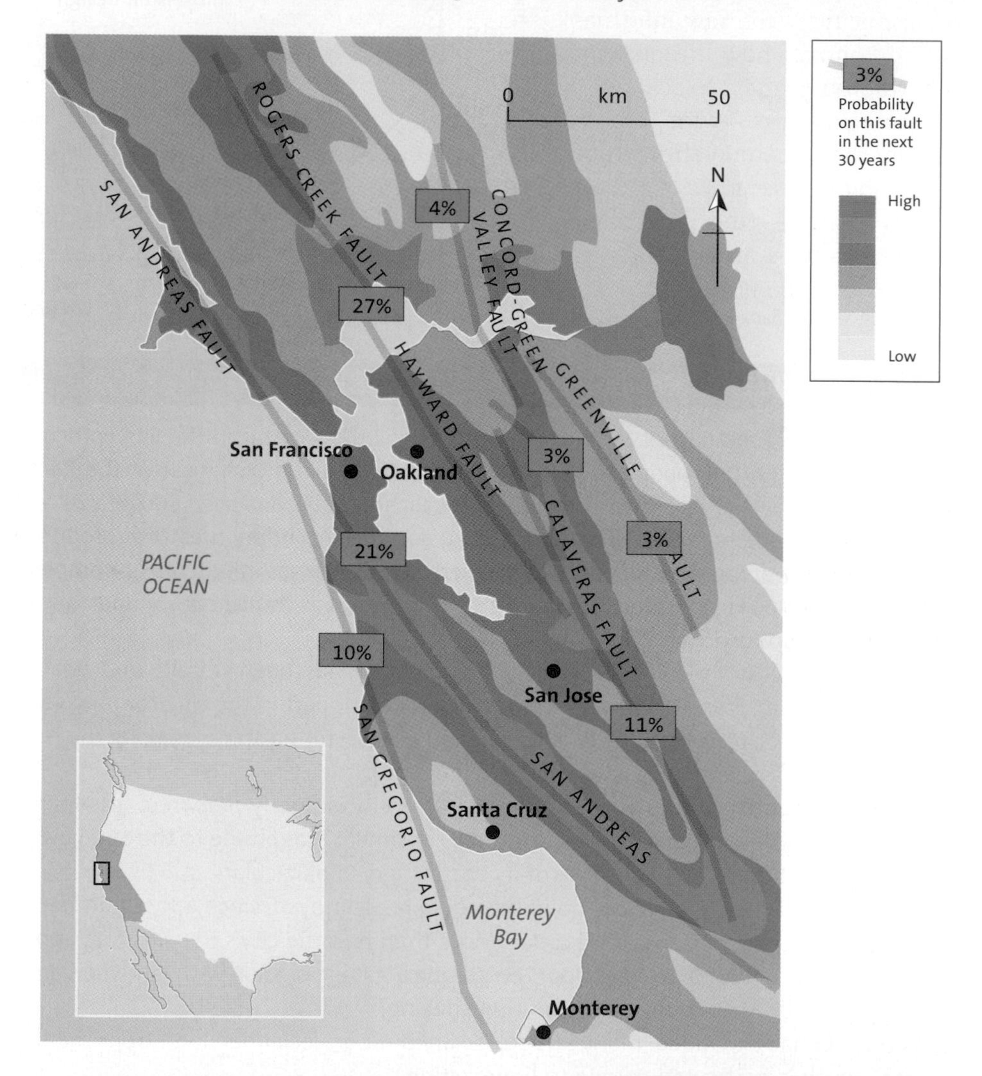

Figure 15
Level of earthquake hazard in the San Francisco Bay Area

Using case studies 3

Question

(a) Is technology the best way to reduce disaster impact from earthquakes?
(b) What is the relationship between earthquake magnitude and scale of losses?
(c) What 'human' factors can cause differences in the geography of an earthquake disaster?

Guidance

(a) Advances in the use of GIS, GPS and satellite technology provide better early warning systems. New building designs and the use of new technologies, combined with better planning enforcement, should mean that new buildings in earthquake-vulnerable areas are better protected. However, a big question remains about the accessibility and affordability of new technologies.

(b) Research a list of major earthquakes, recording number of deaths, damage caused, and magnitude on the Richter scale to assess the question. There is a mild relationship between earthquake magnitude and scale of losses (this is unusual in terms of hazards geography). For example, the 1988 Armenian and 1989 Californian earthquakes were the same size (6.9 on the Richter scale) but 25 000 were killed in Armenia and only 59 in California. The reasons for the differences are:

- geographical variation in the concentration of humans and vulnerability
- geographical variation in the potential for secondary hazards — for example, landslides and fire
- timing of earthquakes — for example, at a busy time of day, or at night

(c) There is a range of factors including population density, remoteness, building and structural vulnerability and extent of preparedness.

Volcanoes

A volcano is a landform that develops around a weakness in the Earth's crust, from which molten **magma**, disrupted from pre-existing volcanic rock, and gases are ejected or **extruded**. The violence of an eruption is determined by the amount of dissolved gases and how easily the gases can escape.

Volcanoes may be classified as:

- the central-vent type, where the eruption takes place from a single pipe
- the fissure type, where magma is extruded along a linear fracture

They may also form low-lying **craters** known as **calderas**, created as a result of subsidence or collapse following an eruption.

Volcanoes may be **active**, **dormant** or **extinct**:

- An extinct volcano is one that is not erupting now and is not likely to erupt in the future.
- A dormant volcano is one that is not erupting now but has erupted since written records were begun (e.g. in the last 200 years), and is likely to do so in the future.
- An active volcano is one that is erupting or is expected to erupt in the near future.

These descriptions are subjective (i.e. they use terms such as 'likely' and 'near future') and should be used with caution.

The distribution of volcanoes is restricted to zones where there is a supply of magma from the Earth's interior. Most volcanoes coincide with the margins of tectonic plates (interplate), but some volcanoes occur above **hotspots** within plate interiors, i.e. intraplate.

The potential impact of a volcano is determined by a number of factors, such as scale (see Table 1), type of eruption (e.g. Hawaiian, Strombolian) and frequency of eruption. Specific factors include eruption duration and length of eruption climax as well as the type of lava, which depends on silica content.

Type of eruption	Volcanic explosivity index (VEI)	Eruption rate (kg s^{-1})	Volume of ejecta (m^3)	Eruption column height (km)	Duration of continuous blasts (h)	Qualitative description
Hawaiian	0 Non-explosive	10^2–10^3	<10^4	0.8–1.5	<1	Effusive
	1 Small	10^3–10^4	10^4–10^6	1.5–2.8	<1	Gentle
Strombolian	2 Moderate	10^4–10^5	10^6–10^7	2.8–5.5	1–6	Explosive
	3 Moderate/large	10^5–10^6	10^7–10^8	5.5–10.5	1–12	Severe
Vulcanian	4 Large	10^6–10^7	10^8–10^9	10.5–17.0	1–>12	Violent
Plinian and ultra-Plinian	5 Very large	10^7–10^8	10^9–10^{10}	17.0–28.0	6–>12	Cataclysmic
	6 Very large	10^8–10^9	10^{10}–10^{11}	28.0–47.0	>12	Paroxysmal
	7 Very large	>10^9	10^{11}–10^{12}	>47.0	>12	Colossal
	8 Very large	—	>10^{12}	—	>12	Terrific

Table 1 *Classification of volcanic eruptions*

The true risk of the volcano is more complex to forecast becauce it is determined by other factors, including:

- population density
- presence or absence of monitoring and alert systems
- infrastructure
- the ability of civil authorities to produce appropriate responses from the local population during a crisis, i.e. community preparedness
- timing of eruption, i.e. day or night

The impacts of a volcanic event vary spatially from a local scale to global. Primary effects include:

- tephra — solid material of varying grain size (volcanic bombs to ash) ejected into the atmosphere
- pyroclastic flows — very hot (c. 800°C), gas-charged, high-velocity flows — a mixture of gases and tephra
- **lava** flows
- volcanic gases including carbon dioxide, carbon monoxide, hydrogen sulphide, sulphur dioxide and chlorine (in 1986, emissions of carbon dioxide from Lake Nyos in Cameroon killed 1700 people)

Secondary effects are:

- lahars — volcanic muds such as those that devastated the Colombian town of Armero after the eruption of Nevado del Ruiz in 1985
- flooding — under-ice volcanic activity can cause catastrophic melting, for example the Grimsvötn glacial burst in Iceland in 1996
- tsunamis — see p. 20
- volcanic landslides
- climate change — the ejection of huge quantities of volcanic debris and gas from large volcanoes (e.g. Krakatau in 1883) can cause a change in average global temperatures

Overcrowding and a shortage of land are forcing increasing numbers of people to live near volcanoes. More than 500 million people — approximately 7% of the Earth's population — live in areas at risk from volcanic hazards. Such land often has richer soils and a better climate than the surrounding region.

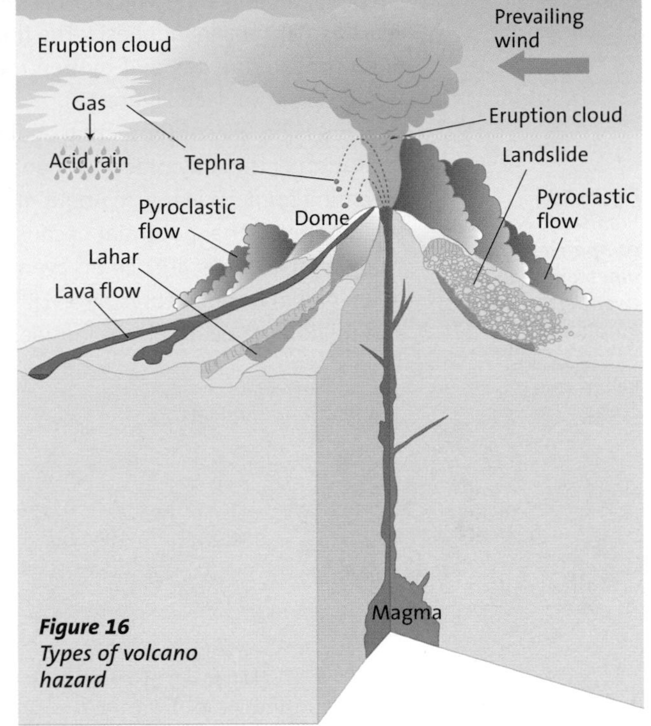

Figure 16
Types of volcano hazard

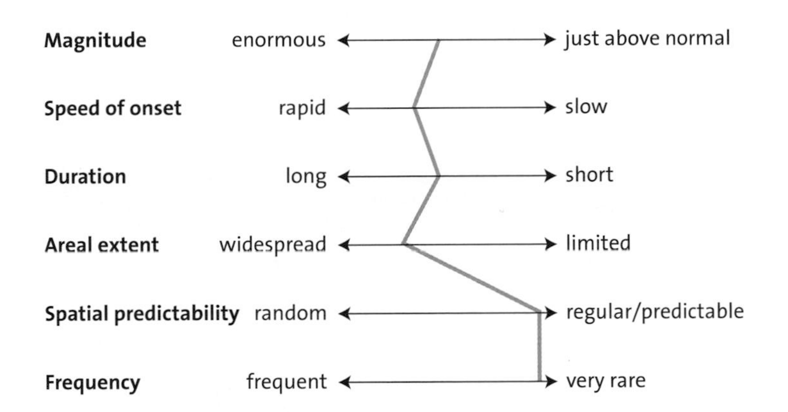

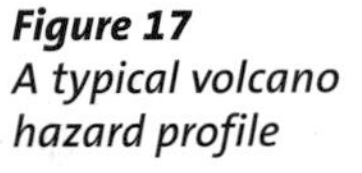

Figure 17
A typical volcano hazard profile

In comparison with other hazards, such as droughts, earthquakes and floods, volcanoes have historically killed far fewer people. Nevertheless they claim a significant number of lives. More than 250 000 people have died in volcanic eruptions in the last 300 years. In any single decade, up to 1 million people may be affected by volcanic activity. This figure is likely to rise as vulnerability increases in populations living close to volcanoes.

SUPERVOLCANO: MYTH OR POSSIBILITY?

Case study 3

According to some scientists, a supervolcano is not only possible but overdue. A supervolcano has the potential to cause global devastation. The impact of such an event would dwarf that of Hurricane Katrina (see p. 43) and the 2004 Asian tsunami (p. 91).

On average, a supervolcano occurs every 50 000 years. The last one struck 75 000 years ago at Toba in Indonesia. The first Yellowstone event (2.1 million years ago) ejected 6000 times more gas and molten rock into the atmosphere than the 1980 Mount St Helens eruption (see Figure 18).

There is no way of protecting against such a rare and catastrophic event — damage limitation is the only practical management option. The issues involved are similar to preparing for a nuclear war. About 40 supervolcano sites are known but Yellowstone has the greatest lethal potential because of its position on a heavily populated continent. The global impacts of such an event centre on emissions of ash and sulphur bringing down global temperatures by 5–15°C. This would lead to failure of the Asian monsoon and millions of deaths from famine.

Figure 18
Prehistoric eruptions compared with nineteenth- and twentieth-century eruptions

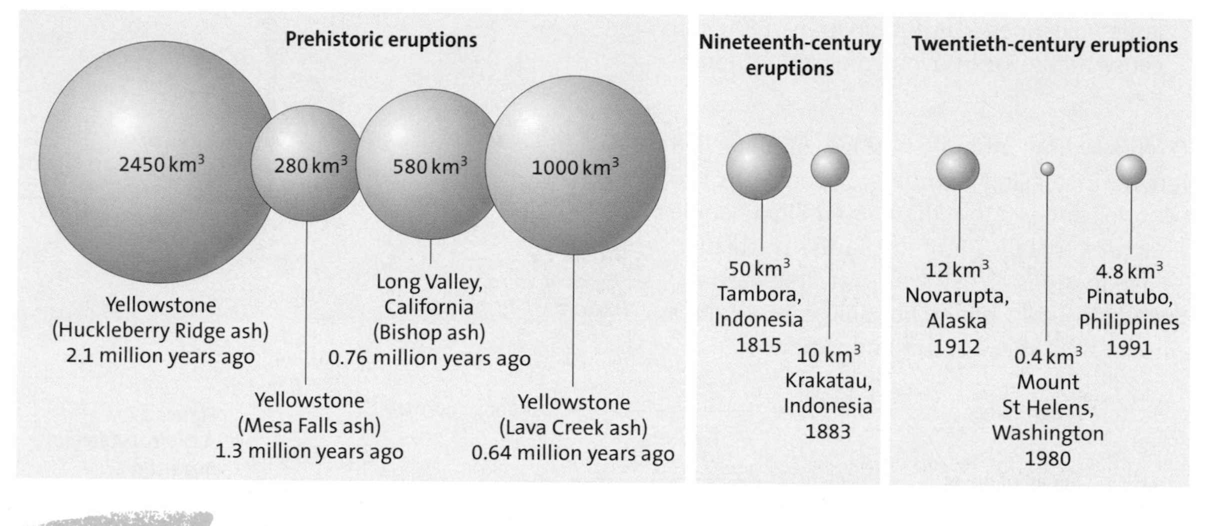

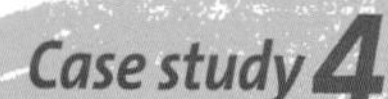

VOLCANIC ERUPTIONS

Mount Merapi: ready and waiting in 2006

Mount Merapi ('Mountain of Fire') is a conical volcano in central Java. It is the most active volcano in the region — it has erupted 68 times since the sixteenth century. Figure 19 shows the location and area at risk.

In April 2006, increased seismicity and an enlarged bulge in the volcano's cone indicated that fresh eruptions were imminent. This posed a particular hazard since the volcano lies close to the city of Yogyakarta and thousands of people inhabit the flanks of Merapi. By early May, active lava flows had begun. On 11 May 2006, with lava flow beginning to be constant, 17 000 people were evacuated from the area. The authorities raised the alert status to its highest level. But many villagers ignored the warnings, saying that they could not leave as they had livestock and crops to tend.

By the end of May 2006, volcanic activity had reduced, despite a 6.2 Richter scale earthquake within 50 km of Merapi on 27 May. There were fears that this could initiate a 'blow' but it did not.

Some volcanic activity continued in June 2006, including lava flows and superheated clouds of gas on the upper slopes. The volcano remains a threat to the people who live nearby. In 2007, Mount Kelud, also on Java, began similar activity.

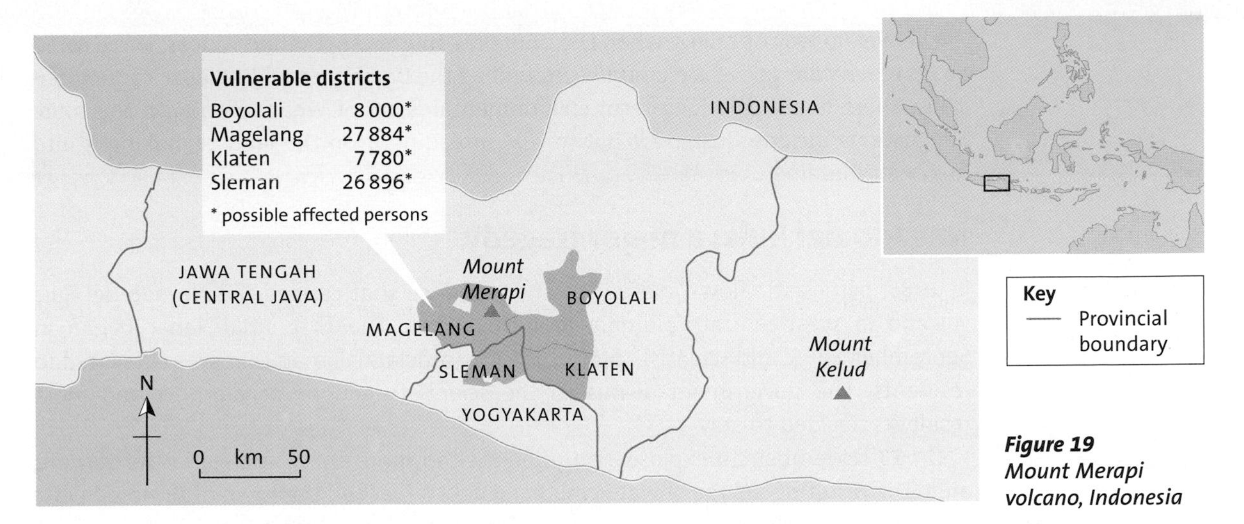

Figure 19
Mount Merapi volcano, Indonesia

Man-made catastrophic mudflow, Lusi, east Java: a quasi-natural hazard

The mud started flowing on 29 May 2006, a few hundred metres from where the gas company PT Lapindo Brantas was drilling an exploratory well nearly 3 km deep. Boiling mud has been gushing up at nearly 50 000 cubic metres a day, equivalent to 20 large baths full every 10 seconds.

Thousands of people have been forced to leave their homes and villages. The mud covers thousands of acres of farmland, roads, schools and factories. The cost of the disaster stands at US$500 million and is rising.

There is speculation over the exact cause of this hazard. The general consensus is that it was human-induced and is related to the drilling operations. One theory suggests that water at high pressure from the 3km borehole forced its way into the surrounding 'muddy' rocks. This water has mixed with the mud material before breaking to the surface, creating a mud volcano (Figure 20).

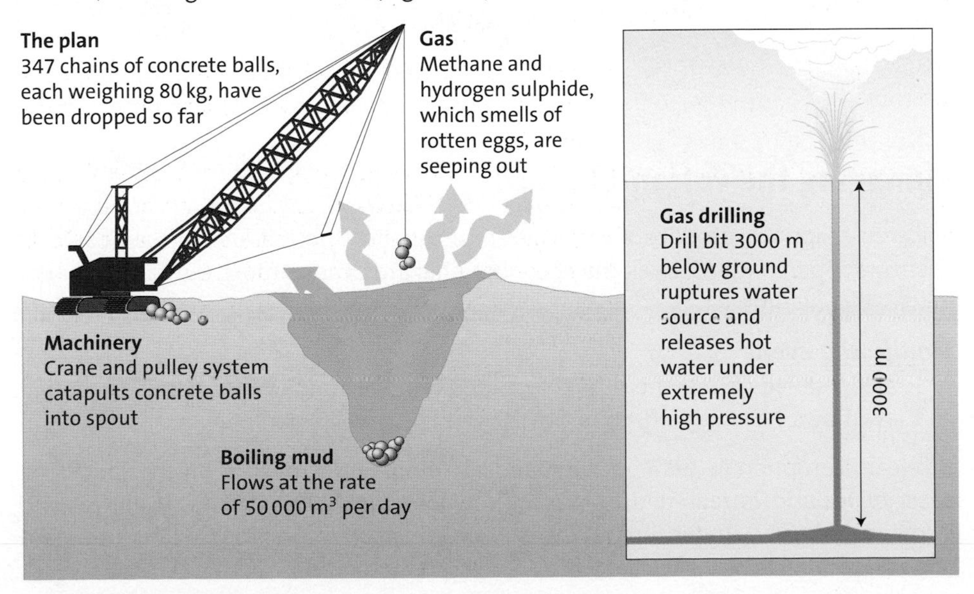

Figure 20
The attempt to stem the flow of mud

There is no way of telling when the mud flow might stop. When it does, there could be grave new danger — the ground surrounding the borehole could collapse catastrophically. There is also the long-term environmental issue of what to do with the mud. Suggestions include flushing it out to sea, spreading it on the land or making it into bricks for building.

Nevado del Ruiz: a major tragedy

In 1985, pyroclastic flows melted the snow and ice that capped the Nevado del Ruiz volcano in west-central Colombia to form lahars. The first small lahar began in September 1985, and scientists produced an unofficial risk map that was circulated to residents. The government dismissed the scientists actions as alarmist and most residents decided to stay.

On 13 November, an explosive eruption melted more snow and ice in the summit area. The resulting deluge of water, mud and debris reached the town of Armero in just 2 hours, overwhelming it in a 40 m wall of mud. Almost 23 000 people died.

Figure 21 *Nevado del Ruiz, Columbia*

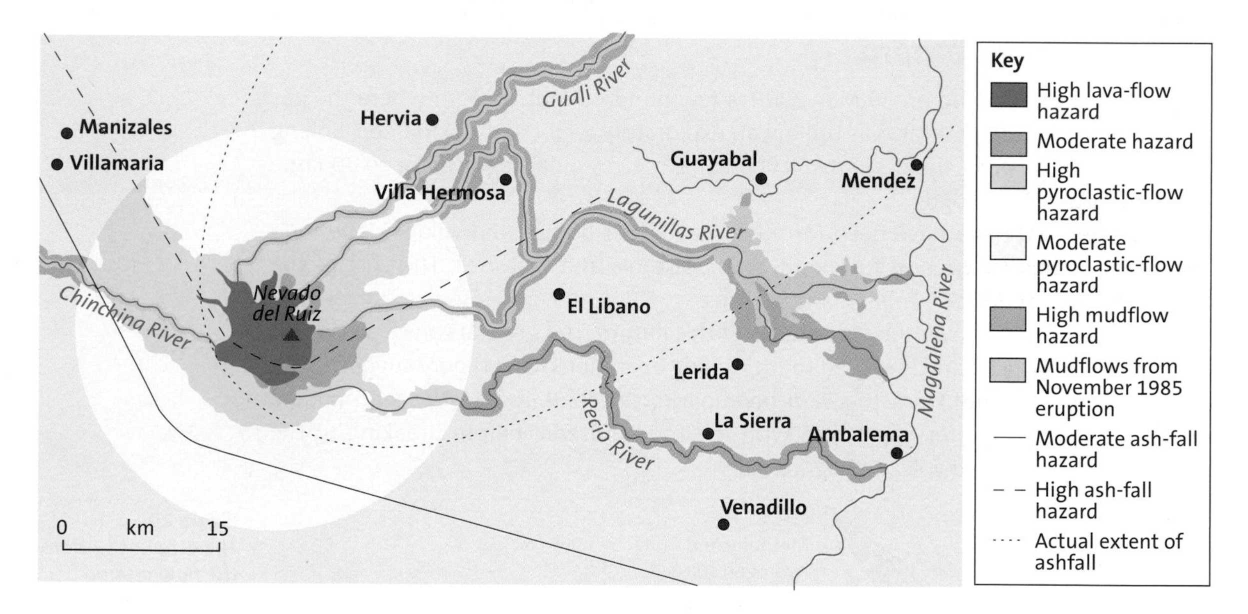

Managing the volcanic hazard

Volcanic eruptions produce a variety of hazards that can kill and injure people or destroy property. Sometimes the secondary hazards cause more damage than the initial volcanic impact.

Modify the event

In some instances it may be possible to modify a volcanic event once the eruption and lava flows have started, either by diverting or chilling the flows.

A volcanic eruption in 1973 on the island of Heimaey (Figure 22), off the southwest coast of Iceland, threatened to destroy a whole community. Seventy homes and farms were buried under tephra and 300 buildings were burned by fires or buried under lava flows.

The lava flow was heading towards the fishing port and harbour — the economic lifeline of the island. If this filled in with lava, the community would not survive. The Icelanders sprayed seawater onto the lava to slow its movement by chilling. More than 30 km of pipe and 43 pumps delivered seawater at a rate of up to 1 m^3 per second. Six million cubic metres of water were pumped onto the flows. The effort saved the port and the residents returned to rebuild their town. The heat from the cooling lava was used to construct a sustainable (geothermal) district heating system.

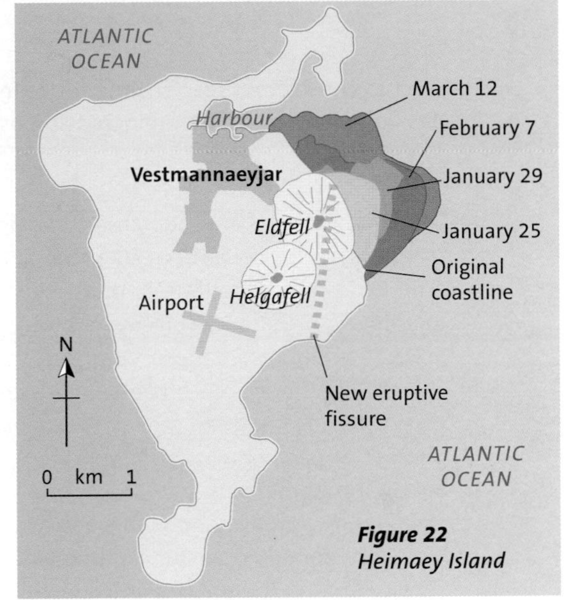

Figure 22
Heimaey Island

Modify vulnerability

Volcanologists have an advantage over seismologists in that volcanoes do not erupt without warning. The warning signs typically take the form of numerous small earthquakes and a swelling of the ground surface, which reflect the passage of magma to the surface. This may be accompanied by gas discharges, which damage vegetation and kill wildlife. However, it is difficult to predict exactly when activity will take place, especially the timing of a major eruption.

Changes in the direction of seismic waves around active volcanoes could help warn of impending eruption almost a year before it occurs. Researchers in New Zealand say they have identified a telltale signature for Mount Ruapehu, which blew its top in 1995, and the finding could be allied to other volcanoes. Seismometers installed around the mountain detect changes in the wave direction as magma pressure builds up in the volcano. This means the 'stress' of the volcano can be measured in situ. It is like looking at a volcano through an X-ray.

Other 'integrated' approaches are now being used to help predict eruptions, using a range of network sensors to make more accurate models (see Figure 23). Monitoring may give time for the area under threat to be evacuated.

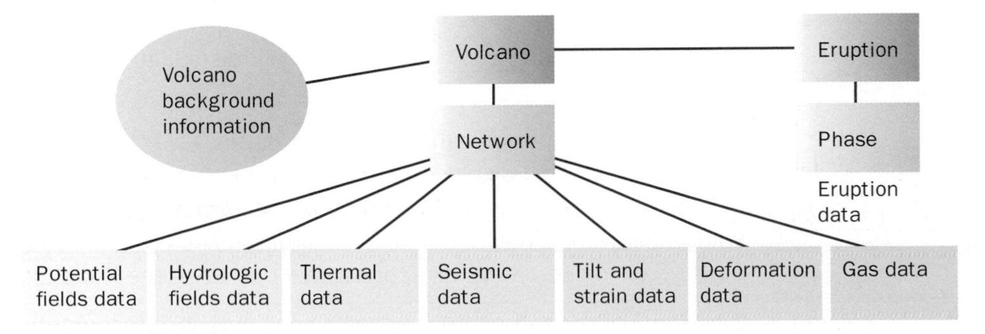

Figure 23
Simplified WOVOdat schema (from the World Organization of Volcano Observatories)

Modify the loss

Insurance policies generally do not cover Earth movements, including shock waves or tremors caused by volcanic eruption. However, most policies provide cover for property loss caused by a volcanic blast, airborne shock waves, ash, dust or lava flow. Fire or explosion resulting from volcanic eruption is usually covered.

Using case studies 4

Question

(a) Discuss the factors responsible for differences in the potential hazard posed by volcanoes.
(b) Has volcanic activity been increasing over recent decades?

Guidance

(a) There are physical and human factors. Physical factors include type of eruption (magnitude, magma type: viscous/runny). Human factors include population density, proximity to eruption, timing of eruption, preparedness etc. Eruptions in areas with limited or no monitoring present the biggest hazard. Eruptions when there is little or no time for evacuation are also dangerous.

(b) The *actual* number has probably not increased, although the *number of reported events* has gone up significantly in the last few centuries. This can be accounted for by increases in the number of people living near volcanoes and the better awareness and increased recording and reporting of events.

Another complication is that the pattern of volcanic activity is not evenly distributed. There are periods of historical time when there are distinct 'highs and lows' in volcanic activity around the globe. For example, Indonesia is currently 'high'.

Tsunamis

Tsunamis are waves initiated by submarine earthquakes, landslides, slumps and volcanic explosions. As such they are secondary hazards. The word tsunami is Japanese, meaning harbour wave. This emphasises the potential impacts on low-lying developed coastal areas.

Most tsunamis consist of a series of waves generated by the rapid movement of the sea bed. They differ from wind-generated waves in a number of ways:

- the wavelength is very long — between 150 km and 1000 km
- velocities may reach 600 km per hour in deep water
- the wave has low amplitude — 0.5–5 m
- the wave height is shallow in relation to the wavelength, making tsunamis almost undetectable in the open ocean

Table 2 *Relationship between tsunami and earthquake magnitudes*

Earthquake magnitude (Richter scale)	Tsunami magnitude	Maximum wave run-up (m)
6.0	Slight	—
6.5	−1	0.5–0.75
7.0	0	1.0–1.5
7.5	1	2.0–3.0
8.0	2	4.0–6.0
8.5	3	8.0–12.0

Source: Smith and Ward (1998)

Earthquakes are a significant cause of tsunamis, especially those that involve vertical displacement at a subduction zone or mid-ocean ridge. Horizontal displacement is not normally associated with tsunamis. The magnitude of a tsunami is related to the height of the wave on reaching the shore (wave run-up — see Table 2).

The shape of the shoreline will have a considerable effect on the degree of impact.

Submarine landslides can also cause tsunamis. Around Hawaii, for instance, 17 major slides have been associated with tsunamis over the last 2 million years.

The distribution of tsunamis (see Table 3) remains fairly predictable in terms of *source* areas. Around 90% of all tsunamis are generated within the Pacific Basin, associated with activity at plate margins. Most are generated at convergent plate boundaries where subduction takes place, particularly off the Japan–Taiwan island arc (25% of all events), South America and the Aleutian Islands area. Geological records indicate that huge tsunamis have affected other areas, for example the Santorini eruption around 1450 BC (Crete), which impacted on the Mediterranean basin, and the Storegga Slide off the coast of Norway around 5250 BC.

Body of water	Number of tsunamis
Pacific Ocean	828
Indian Ocean	65
Atlantic Ocean (including Caribbean)	61
European seas	60
Other inland seas and lakes	15
Southern Ocean	1
Arctic Ocean	0
Unidentifiable	71
Source: NOAA	

***Table 3** Distribution of tsunamis*

The impact of a tsunami will depend on a number of physical and human factors:

- the height of the waves generated and the distance they have travelled
- the duration of the event
- the physical geography of the coast, both offshore and at the shoreline
- the presence or absence of ecosystem protection by coral reefs and mangroves
- the coastal land use and population density
- the quality of early warning systems and the time between warning and impact
- the timing of the event, i.e. night or day and season, especially in tourist coastal resorts

***Figure 24** Tsunami events since 1900*

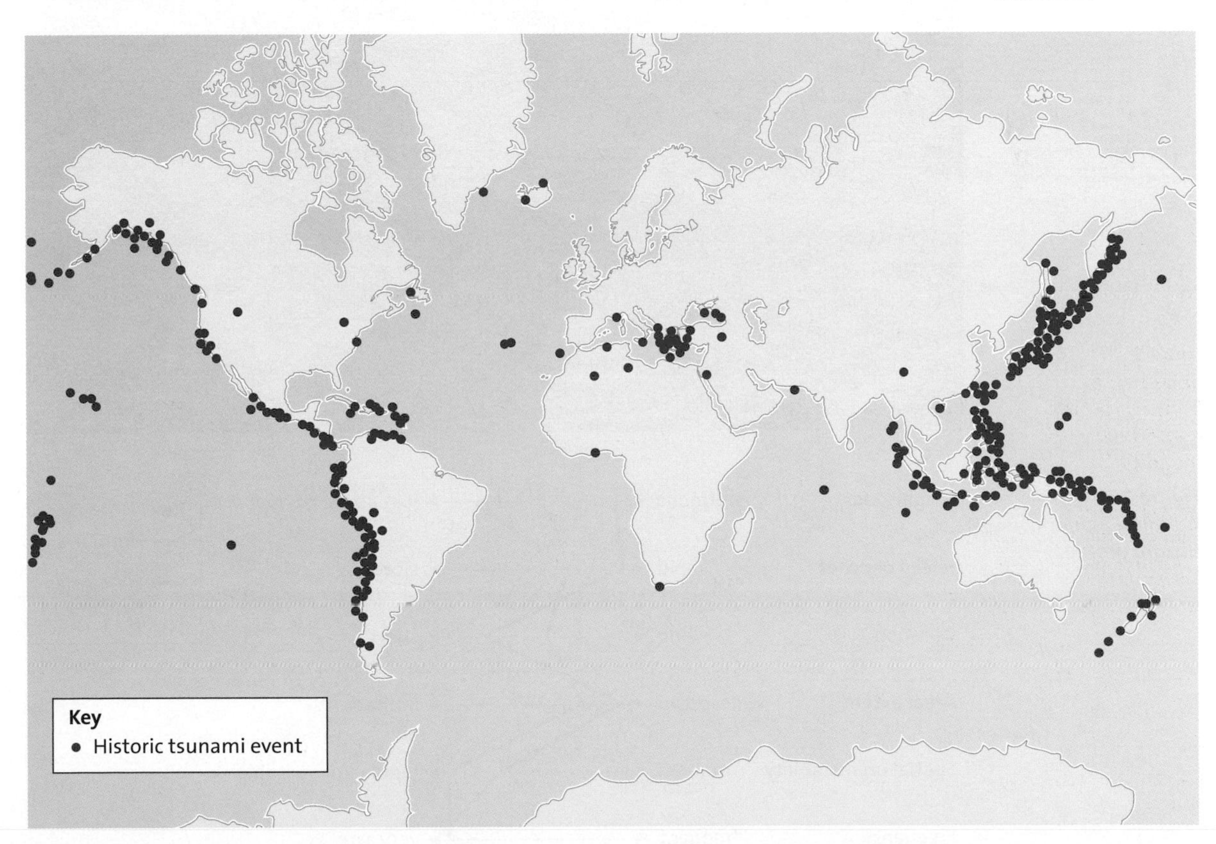

Any large tsunami will have a range of social and economic consequences (see Table 4 and Figure 25).

The 2004 Asian tsunami is discussed on p. 91.

Table 4 *Consequences of tsunamis*

Human/social	Economic
Loss of life	Loss of houses and assets
Children left with no parents; widows	Destruction of fishing vessels
Damage to education facilities (schools, universities)	Salination of agricultural land and loss of crops
People injured	Loss of tourism (and related infrastructure, e.g. hotels)
Displacement and forced migration	Damage to other businesses

Figure 25 *The aftermath of the 2004 Asian tsunami near the coast of Sumatra in southeast Asia*

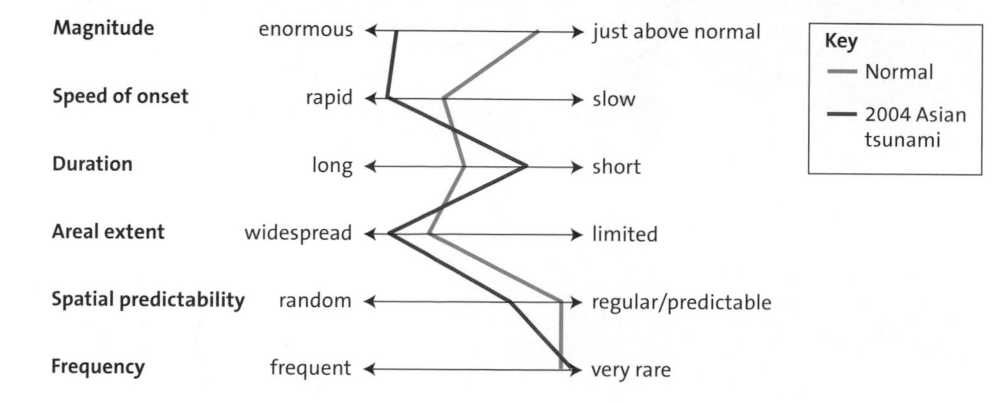

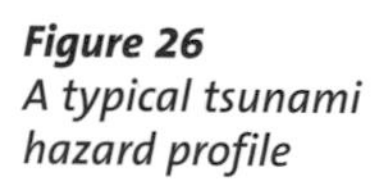

Figure 26 *A typical tsunami hazard profile*

TSUNAMI IN THE BRISTOL CHANNEL?

Flooding occurred in the Bristol Channel at around 9 a.m. on 20 January 1606 (in the modern calendar this is 30 January 1607). The event is recorded on plaques at churches in Kingston Seymour in Somerset and at Goldcliff, St Brides, Redwick and Peterstone in Monmouthshire. Eyewitness accounts tell of 'huge and mighty hills of water' advancing at a speed 'faster than a greyhound can run' and only receding 10 days later. The total death toll has been estimated to be around 2000.

In the open sea between north Devon and Pembrokeshire, the wave was just under 4 m (13 ft) high. As it entered the constricting funnel-shaped Bristol Channel and Severn Estuary, the wave height increased to 5 m (16 ft) along the Glamorgan coast, 5.5 m (18 ft) along the Somerset coast, and over 7.5 m (25 ft) by the time it reached the Monmouthshire coast.

The velocity of a tsunami is related to its height, so as it moved up the estuary and was squeezed between the opposing shores of England and Wales, it became faster, striking the coast at just over $12\,m\,s^{-1}$ (27 mph) in north Devon and southwest Wales, to just under $14\,m\,s^{-1}$ (31 mph) along the Glamorgan coast, to $14.5\,m\,s^{-1}$ (32 mph) in Somerset, and over $17\,m\,s^{-1}$ (38 mph) in Monmouthshire.

The tsunami penetrated a considerable distance inland on the flat coastal areas. The maximum inland penetration in north Devon and southwest Wales would have been just under 2.5 km (1.55 miles), in Glamorgan just over 3 km (1.86 miles), in Somerset just under 4 km (2.5 miles), and in Monmouthshire just under 6 km (3.7 miles). The flood-waters reached further inland in some places, such as to the foot of Glastonbury Tor (14 miles inland), because the land surface slopes landward in many coastal wetland areas, so once the wave collapsed, the water flowed landward under gravity rather than back to the sea.

The cause of the tsunami is uncertain but possibilities include a landslide off the continental shelf between Ireland and Cornwall, or an earthquake along an active fault system in the sea south of Ireland. This fault system has experienced an earthquake of more than 4 on the Richter scale within the last 20 years, so a bigger tsunami earthquake is a possibility. It may have been that an earthquake triggered a submarine slide.

The evidence for the 1607 flood/tsunami is incomplete and should be treated with a degree of caution. However, a repeat **storm surge** of this magnitude would be the UK's costliest natural disaster. A risk management company has calculated that a similar event in the Bristol Channel could cost £13 billion at 2007 prices.

Managing the tsunami hazard

Modify the event

Tsunamis are among the world's most terrifying hazards. Modification of the event is impossible — there are no technologies for either prevention of the tectonic disturbance or diversion or deflection of the wave itself. However, research is beginning to show that replanting of coasts may be a way of affording better protection and therefore modification of the event.

The Asian tsunami of 2004 was undoubtedly an irresistible force of nature, but fewer people might have died if coasts had been protected by mangroves or other types of dense coastal forest.

There is still considerable debate as to the effectiveness of these so-called buffer zones. Mangroves are known to be effective at dissipating energy from waves whipped up by the wind. Modelling studies also suggest that shore vegetation can reduce the flow speed and height of an oncoming tsunami, but there is limited field evidence to back this claim.

Many scientists and conservationists are happy to encourage replanting of the coastal strip because it offers a more natural and diverse habitat than deforested regions and coastal development for tourism.

Modify vulnerability

The monitoring of earthquake and volcanic activity should enable warnings to be given to populations at risk, so that evacuation can take place. A centre has been established on Hawaii (the Pacific Warning System) that will give warnings to countries situated around the Pacific Rim. This station is linked to several seismic and tidal observatories throughout the region and also to satellite information provided by the National Oceanic and Atmospheric Administration (NOAA) — see Figure 27. A similar system now operates in the Indian Ocean as a result of the 2004 Asian tsunami, much of it financed by Japan and India.

Local and indigenous knowledge captured in legends and songs may also save lives. The basic message is: if the ground shakes, the sea recedes and you hear an

Figure 27
The tsunami warning system

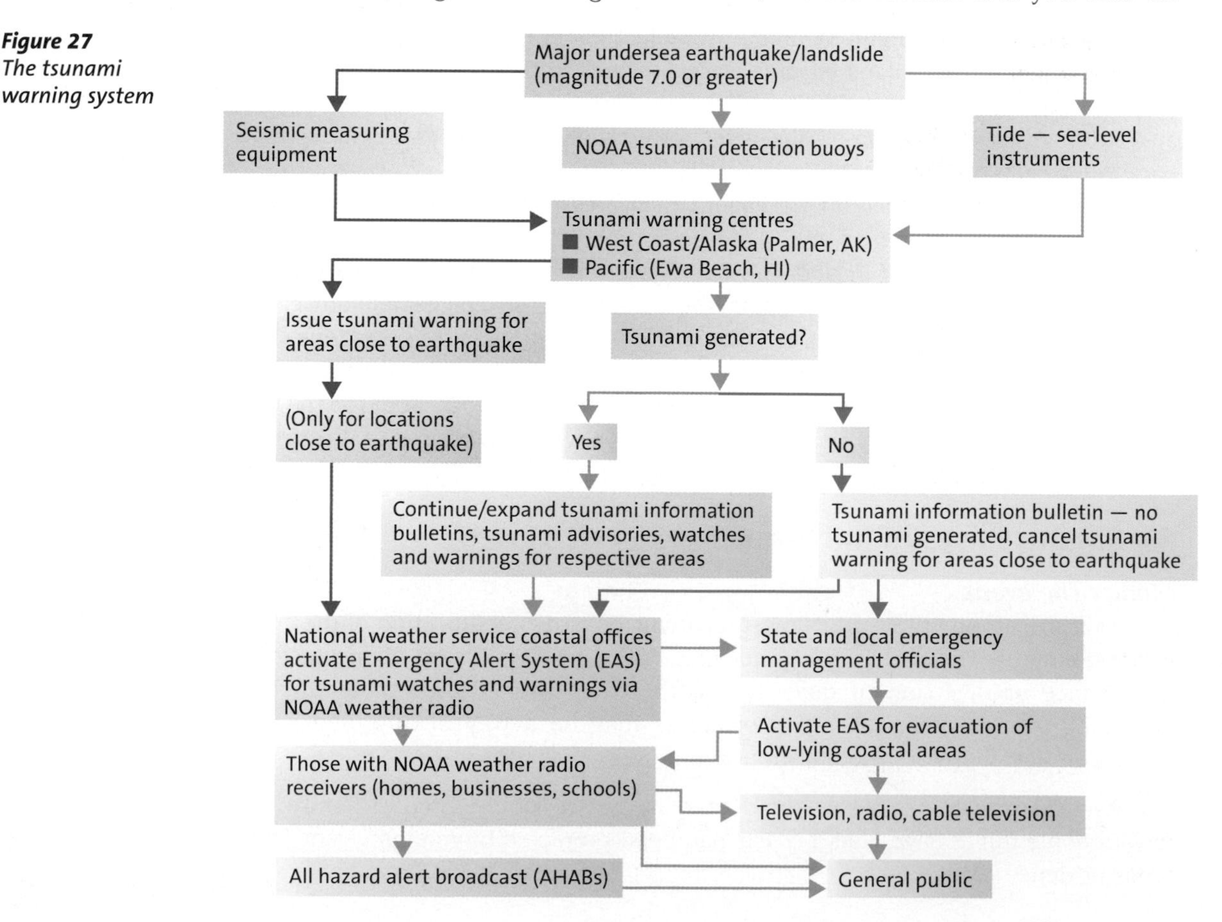

approaching roar, then flee to higher ground. If you are already on the water, take your boat out to sea as tsunami waves grow taller in shallower water.

Other ways to modify vulnerability to the tsunami hazard include land-use planning (this requires political control) and specialist hard engineering and building design approaches.

Modify the loss

The loss of life resulting from the Asian tsunami in December 2004 was enormous (just under 300 000) and the economic losses were large (approximately US$10 billion), but the insurance losses were comparatively modest. Few people had personal life and property/possession insurance and no major airports, ports or industrial complexes were hit. Most of the claims arose from damage to tourist resorts, fishing and interruption to businesses.

For example, the final insurance payout to affected areas in Thailand will not reflect the full extent of the damage because many small businesses were not insured or did not have adequate cover. This is an ongoing problem in many of the poorest parts of the world.

5 *Using case studies*

Question

'Systems of prediction and risk analysis are well developed so catastrophe is avoidable. The problem in developing countries often comes down to making difficult development choices from among the many competing demands.'

Source: World Bank, *Lessons from Natural Disasters* (2005)

(a) In the case of a tsunami, is 'catastrophe always avoidable'?
(b) What are the other competing demands that are indicated?
(c) Why does relocation of entire villages away from the coastal strip 'hazards' present new challenges?

Guidance

(a) Even with well-developed monitoring systems, if the tectonic shock is close to a populated area then the resulting wave may arrive in a matter of minutes. This means that there will be little time for early warning systems to take effect and no time to deliver warnings or carry out evacuations.

(b) There is a range of competing demands from limited financial resources. These can include infrastructure, health and welfare systems, sanitation, housing and education. Tsunami disaster mitigation is a periodic need rather than a constant one and this adds to the difficulty of prioritisation, especially once the disaster has fallen out of the international media and the *immediate* relief needs have been met.

(c) When relocating people away from one risk, it is important to keep exposure to new risks in mind. When people are moved away from coastal zones, the tendency to return is almost irresistible. While it may be important to settle people away from tsunami-prone areas, in situ reconstruction should be promoted after earthquakes to take advantage of existing infrastructure and community facilities, while minimising resettlement and its attendant social dislocation. In situ reconstruction stimulates self-help efforts in low-cost reconstruction. It also provides the opportunity to build on knowledge from the experiences of other developing countries as they face similar emergencies.

Part 3

Hydrometeorologic hazards

Hazards related to the weather, climate and water are among the most serious hazards affecting human populations. Compared with earthquakes and volcanoes, climatic hazards can appear unspectacular but they often have a more significant impact. Table 5 shows the total number of people reported killed by hazard type and location over the period 1996–2005.

Table 5
Number of people killed by hazard type and location, 1996–2005

Levels of development				
Hazard type	**High**	**Medium**	**Low**	**Total**
Avalanches and landslides	365	6 953	546	7 864
Drought and famine	0	842	220 879	221 721
Earthquakes and tsunamis	2 625	306 845	82 140	391 610[a]
Extreme temperatures	48 235	11 041	973	60 249
Floods	3 471	73 490	13 276	90 237
Forest fires	197	259	4	460
Volcanic eruptions	52	10	200	262
Windstorms	5 813	51 411	5 186	62 410
Subtotal hydrometeorological	58 081*	144 447*	240 864*	443 392*
Subtotal geophysical	2 677*	306 855*	83 340*	392 872*
Total natural disasters	60 758*	451 302*	323 204*	835 264*

[a]Note distortion caused by the 2004 Asian tsunami and the Kashmiri earthquake.

Drought and famine, extreme temperatures, floods and windstorms can be classified as atmospheric/hydrologic. 'Other' non-classified natural disasters are not reported in the table, but are included in the totals indicated with an asterisk (*).

Source: World Disaster Report 2006, International Red Cross

Hazards can be defined by their meteorological causes and character and this is the classification we will use here. An alternative approach is classification by impact intensity, measured by the number of people affected. In many respects this is a more useful measure, but it suffers the problem that hazard impact may vary with culture, wealth, expectations etc. The impact is strongly controlled by the relationship of the physical event to human vulnerability (see Figure 28).

Impact frequency is a useful classification for a given region as it helps emergency services to forward-plan. However, accurate long-term records for the frequency of hazards do not exist for all parts of the world. The more frequently a hazard occurs, the better adapted people might become, and therefore the risk will be reduced.

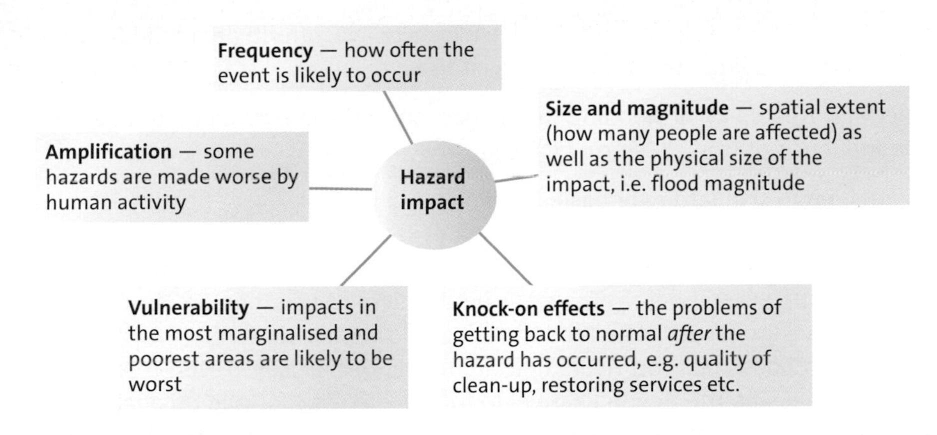

Figure 28
Hazard impact

An example is the building of river flood defences in the UK to protect people and property from the frequent flood hazards on the River Severn.

Figure 29
Barriers installed by the Environment Agency on the River Severn in Shrewsbury in 2004

River floods

Flooding is evident in more than one-third of the world's land area, in which 82% of the world's population resides. Some of the most flood-prone regions include large areas of the mid-western USA, Central America, coastal South America, Europe, eastern Africa, northeast India and Bangladesh, China, the Korean peninsula, south-east Asia, Indonesia and the Philippines.

Large areas of China, such as the Yangtze River basin, are subject to significant flood risk, which affects large areas. This is consistent with the idea that flood-prone areas are also areas of intensive agricultural production, development and large populations. River valleys are attractive places to live, offering flat land for a variety of industrial, residential and commercial uses.

The high frequency of flooding in Bangladesh and the surrounding areas reflects the influence of tropical storms (cyclones).

In addition to standard river basin flooding (usually single-event flooding), a number of other types of flood can be classified:

- Flash floods occur with little or no warning, on timescales ranging from several seconds to several hours. They can be deadly because they are associated with rapid rises in normal flow levels and high flow velocities.
- Ice-jam floods occur on rivers that are totally or partially frozen A rise in water level breaks up the frozen channel to form ice flows that pile up against bridges and other potential obstructions. The jammed ice creates a dam, which can cause devastation when the blockage is released.
- Dam- and levee-failure floods are relatively uncommon but may be associated with inadequate maintenance or unusually high precipitation. In 2007, the Ulley reservoir in northern England, which was built in the 1870s and is now only used for recreation, had to be drained in a fight against time to avoid catastrophic flooding of a wide area of south Yorkshire. Cracks appeared in the dam after torrential rain and it was feared the dam would collapse. The threat led to the closure of the M1 motorway for several days.

Figure 30
An aerial view of Ulley Reservoir after heavy rain in June 2007, which caused severe flooding in south Yorkshire

Reuters/Owen Humphreys

These common types of flooding are frequently interlinked, as shown in the southwest Midlands floods in June–July 2007 (Figure 31).

Causes of river flooding include:

- excessive rainfall resulting from atmospheric processes such as monsoonal rain, tropical cyclones or a series of mid-latitude depressions; much of the flooding in the UK in 2007 was associated with a series of depressions dropping considerable rainfall on already saturated ground (see *Case study 6*)

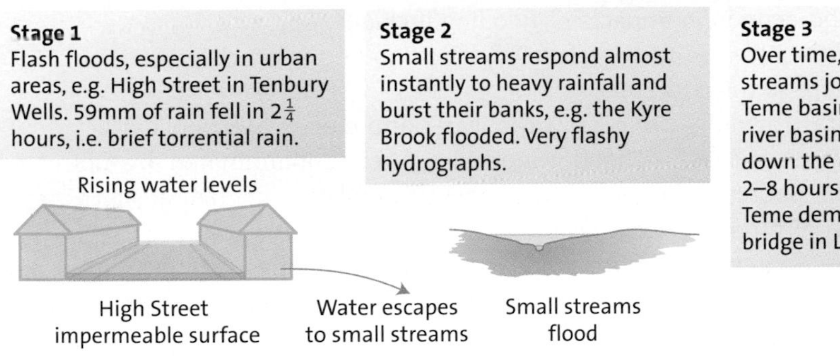

Figure 31
A southwest Midlands' flood model

- intense rainfall events (storms) typically resulting in flash floods, but may affect only a relatively small area
- rapid snowmelt in late spring and early summer; impacts can be devastating
- volcanic eruptions — the extreme heat created by magma can result in the lifting and melting of ice (e.g. Grimsvötn, Iceland in 1996)
- landslides collapsing into reservoirs, causing failures — a potential problem associated with the Three Gorges Dam

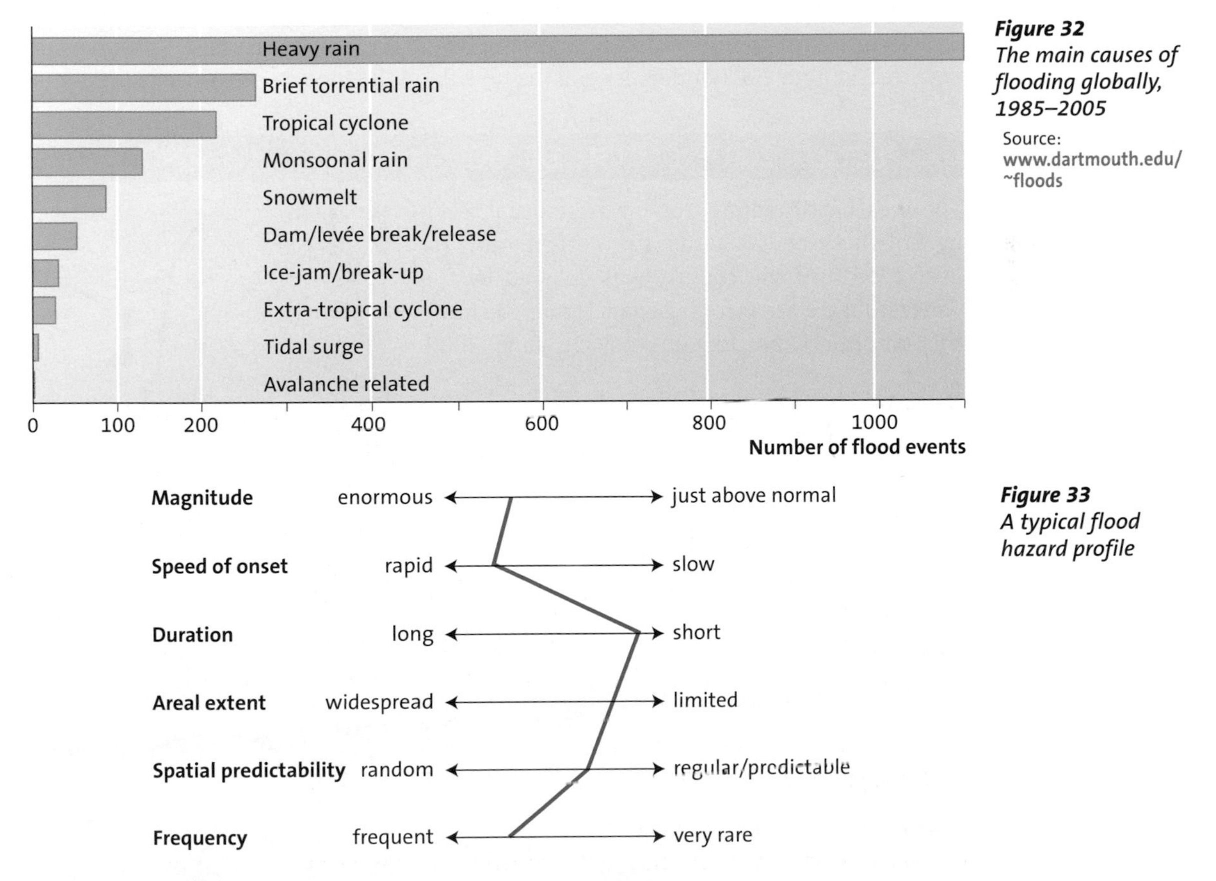

Figure 32
The main causes of flooding globally, 1985–2005

Source: www.dartmouth.edu/~floods

Figure 33
A typical flood hazard profile

The initial impact of flooding is inundation by water of areas not normally submerged. Flood impacts usually become significant only when there is damage to property, travel inconvenience or loss of personal assets.

The social and economic impacts of flooding include:

- deaths and injuries — the death toll due to flooding is much greater in LEDCs than in more developed areas
- loss of infrastructure such as railways, water and electricity supplies
- pollution and disease — flood waters are often contaminated by untreated sewage; the poorest societies are the most vulnerable
- inundation of properties and businesses by water
- loss of economic income from tourism and other activities (e.g. farming)
- increased insurance premiums (*Case study 6*)

The great flood of the US midwest in 1993 was arguably the costliest flood in history. Damages were estimated at US$20 billion; however, only around 50 lives were lost. The Mississippi River spread across 352 counties in nine states and displaced at least 50 000 people. Over 5 million hectares were inundated — this again was a flood of regional scale.

The localised flash flooding at Boscastle on 16 August 2004 saw peak river flows estimated at around 140 cumecs (m^3 per second). This was a 1 in 400 year event. Since then, efforts to rebuild the village's infrastructure and to implement a flood scheme have involved moving a bridge and widening the river, as well as various streetscape improvements to enhance the character of the village. A second flood occurred in June 2007 and in general the new flood defence works proved successful.

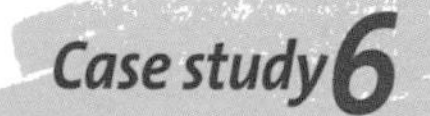

UK FLOODS, 2007: 'DOUBLE DELUGE'

The weather in spring 2007 was unusual, even by British standards. After the hottest April for over a century, the country was deluged by the wettest June since records began. Figure 34 shows the synoptic chart for Saturday 23 June 2007.

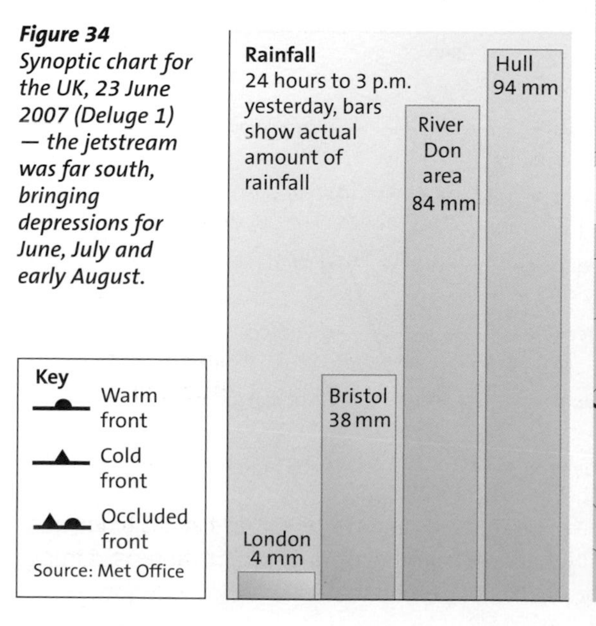

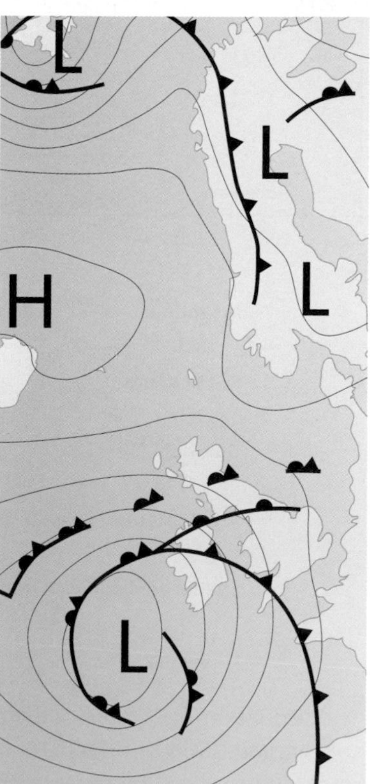

***Figure 34** Synoptic chart for the UK, 23 June 2007 (Deluge 1) — the jetstream was far south, bringing depressions for June, July and early August.*

Figure 35 *Tewkesbury, Gloucestershire was badly affected by flooding in July 2007 (Deluge 2)*

This rainfall event was devastating because the ground was already saturated and the rivers high. Seven people died in the subsequent floods; 27 000 homes and 5000 businesses were evacuated. The final bill was expected to be in excess of £2.5 billion.

The June floods were followed by further flooding in the south Midlands (particularly in Gloucestershire and Oxfordshire) in July. The damage was again estimated at around £2.5 billion insured losses, higher than the 'great storm' of 1987 (see *Case study 12*, p. 46).

The 'double deluge' raises significant questions. First, although the June/July rainfall was a 1-in-200-year event, a combination of climate change and human factors, such as urbanisation, building on floodplains, and changing land use, may make such an event a more common occurrence (every 20–30 years by 2080); the responsibility of the government to provide adequate flood defences will increase. The Environment Agency says it will need to spend £750 million per year between 2008 and 2011 to protect areas threatened by floods. Second, there are the issues of insurance, which may prove too costly, or be refused for areas perceived as at risk, without householders taking extra precautions by providing flood protection (skirts) and redesigning their houses to be floodproof.

THE WORST FLOOD EVER?

The Huang He (Yellow) River (Figure 36) is prone to flooding because of the broad expanse of plain that lies around it. One of the major reasons for the flooding is the high silt content, which gives the river its yellow tint (and thus its name). The silt constitutes as much as 60% of the volume, and builds up until the river is actually higher than the surrounding land.

The tendency to flood is exacerbated by ice dams, which block the river in its upper course; the dams back up the water and when they break release devastating walls of water.

The worst flood occurred in 1931. Between 1 million and 3.7 million people died. The death toll was made much higher by the resulting famine.

The history of flooding has prompted the communist Chinese government to embark on a programme of building dams for flood control. The dams, however, have not proven entirely effective and have been the target of criticism from environmentalists.

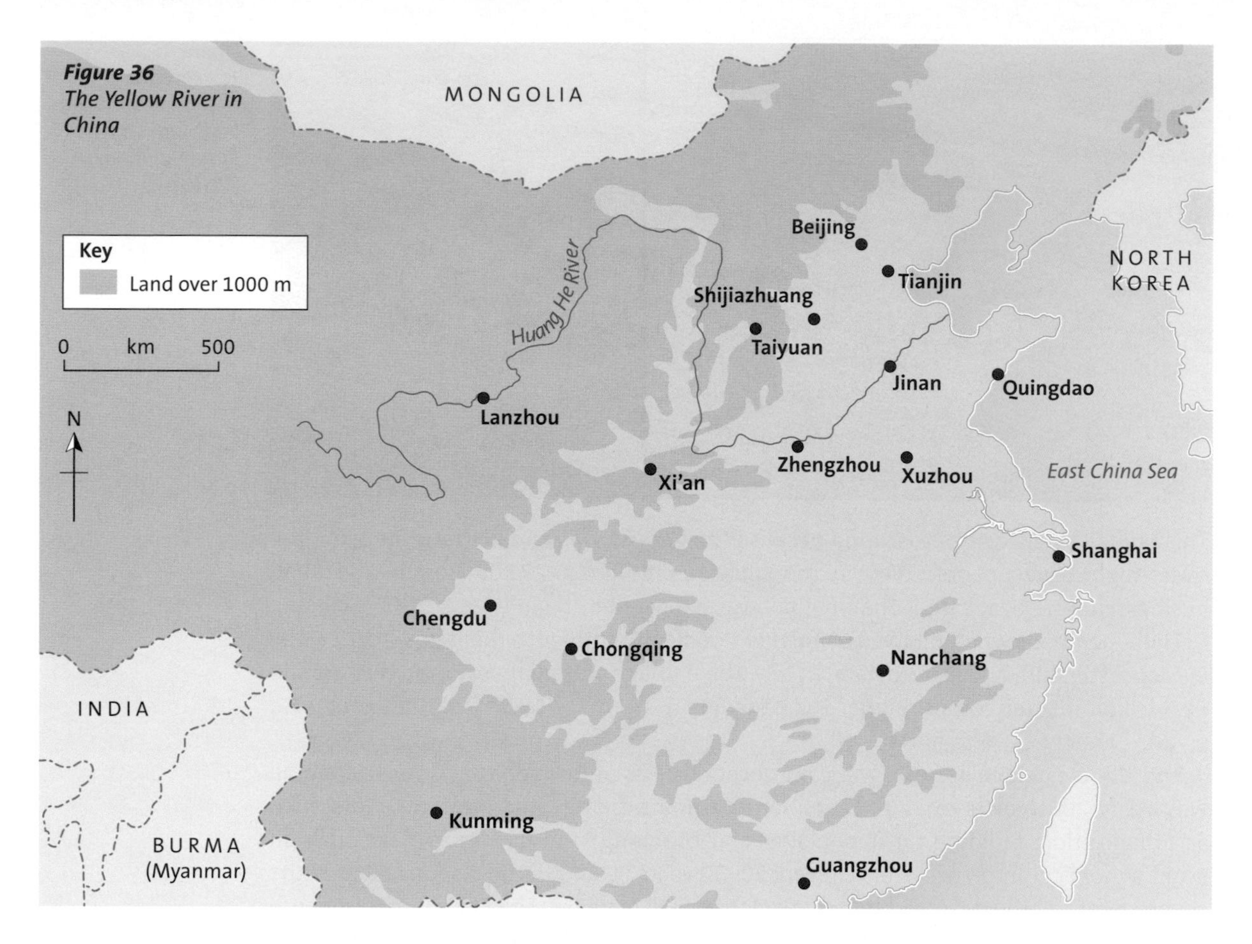

Figure 36
The Yellow River in China

Managing the flood hazard

Modify the event

The root causes of flooding (long-duration and/or high-intensity precipitation/snow-melt) cannot be prevented easily, but various hard and soft engineering options are available to address the flood hazard:

- Flood abatement measures — for example, reafforestation and revegetation of river banks, which should improve levels of interception and reduce overland flow.
- Storing water through building dams and other storage reservoirs to control discharge of water.
- Channel and bank modifications — reducing the bending (sinuosity) of a channel and increasing its hydraulic radius (efficiency) will mean water can be moved more quickly.

Modify vulnerability

Effective flood forecasting depends on monitoring water flow in areas at risk. Such systems are essential for minimising loss of life and property damage. Most developed countries now have dedicated centres or agencies to monitor dangerous river conditions. In England and Wales, the Environment Agency monitors flood risk 24 hours a day. Information is available on a dedicated telephone line or on its website **www.environment–agency.gov.uk/subjects/flood**.

D. Holmes

Figure 37 *Boscastle has become 'famous' for its flood, and visitor numbers have increased by 30%*

Many developing countries still have little in the way of forecasting infrastructure, and effective flood warnings are comparatively rare. This is a major problem as the chance of saving lives rises in proportion to the length of the warning time.

In the Netherlands, floating houses have been designed to slide up and down on steel rails. A more cost-effective alternative is to put garaging downstairs with living accommodation upstairs, and to design flood-proof electricity systems.

Modify the loss

Insurance to cover the cost of flood damage is an important part of flood protection. However, the increasing number of houses being built on floodplains is a cause for concern. Premiums are becoming too expensive to make them reasonably affordable for individual housholders (see *Case study 6*).

6 Using case studies

Question

(a) 'England has a penchant for building houses in silly places...some £200 billion worth of houses, business and other infrastructure is at risk from flooding.' (*The Economist*, July 2007) Why are houses built in 'risky' places, and how can flood risk be taken more seriously?

(b) How and why can channel realignment schemes have undesired impacts further downstream?

Guidance

(a) New houses are needed to accommodate a growing population, especially in the southeast. Often the floodplain is the only land available. Much of the remaining land is protected (green belt). Planning laws have been tightened and the Environment Agency now has more powers. Insurance may end up being a bigger constraint on the house builder than planning laws.

(b) Modification of the channel speeds up water flow. This can lead to increased flow downstream, even in areas that previously did not experience risk of flooding. This presents a particular difficulty because management of the river for flood requires a whole-catchment approach, which is costly.

Droughts

Droughts are difficult to define. They are usually associated with a long, continuous period with little or no precipitation. Drought is not a purely physical phenomenon, but rather an interplay between natural water availability and human demands for water supply. Drought ranks as the natural hazard with the greatest negative impact on human livelihood as it affects millions of people. It is a slow-onset hazard, i.e. it does not directly lead to many deaths, but indirectly, as a result of the subsequent famine, many thousands of people die.

Andrew Holt/Alamy

Figure 38
Cattle herders on parched land in Ethiopia

Three types of drought are commonly identified:

- **meteorological drought** — when there is a prolonged period with less than average precipitation
- **agricultural drought** — when there is insufficient moisture for average crop or range production; this condition can arise even in times of average precipitation, due to soil conditions or agricultural techniques
- **hydrologic drought** — when the water reserves available in sources such as aquifers, lakes and reservoirs falls below the statistical average; this condition can arise even in times of average (or above average) precipitation when increased usage of water diminishes the reserves

Drought can occur virtually anywhere as its causes are widespread. It can affect both the interior and coastal regions of most continents. The most drought-prone areas include parts of:

- western and the midwest USA
- central America and northeastern Brazil
- the sub-Saharan belt and the Horn of Africa

- southern and central Africa and Madagascar
- southern Spain and Portugal
- central Asia, northwest India and northeast China
- southeast Asia, Indonesia and southern Australia

About 38% of the world's land area has a level of drought exposure. This 38% contains around 70% of both the total population and the agricultural value produced. Some areas experience a 'crisis cycle' of deadly drought and floods. Ethiopia's arid south experienced this crisis cycle in 1984 and 2006. Floods followed months of drought, hitting the population when it was most vulnerable. In 2006, approximately 280 000 subsistence farmers were displaced from their villages into makeshift shelters as seasonal rivers switched from being dustbowls into raging torrents.

Direct impacts of drought include reduced cropping, rangeland and forest productivity, increased fire hazard, increased death rates among livestock and wildlife, and damage to wildlife and fish habitats. A reduction in crop productivity usually results in less income for farmers, increased food prices, unemployment and migration. The impacts can be classified as economic, social and environmental (Table 6).

Table 6
Direct and indirect impacts of drought

Economic	Social	Environmental
Loss of national economic growth, slowing down of economic development	Loss of human life from food shortages, heat, suicides, violence	Increased desertification
Damage to crop quality, less food production	Water user conflicts	Damage to animal species
Loss of productivity from dairy and livestock production	Public dissatisfaction with government regarding drought response	Reduction and degradation of fish and wildlife habitats
Insect infestation	Reduced quality of life, which leads to changes in lifestyle	Increased stress to endangered species
Loss of hydroelectric power	Population migrations	Wind and water erosion of soils
Range fires and wildfires	Mental and physical stress	Damage to plant species
Damage to fish habitat, loss from fishery production	Increased poverty	Lack of food and drinking water

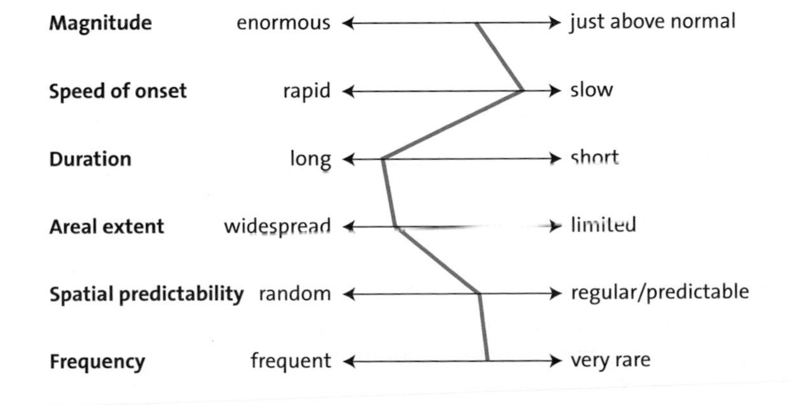

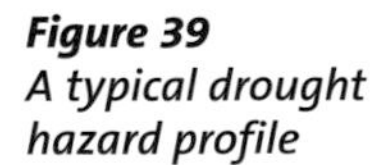

Figure 39
A typical drought hazard profile

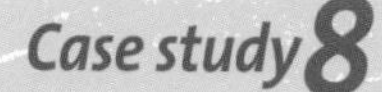

Case study 8 — THE AUSTRALIAN DROUGHT, 2006

Australia has one of the most variable rainfall climates in the world. The 2006 drought was the worst for 1000 years and is on-going. Over the long term, about 3 good years of rain and 3 bad years out of 10 can be expected. These fluctuations have many causes but the strongest is the climate phenomenon called the Southern Oscillation. This is a major air pressure shift between the Asian and east Pacific regions. Its best-known extreme is El Niño.

For the 12-month period from July 2006 to June 2007, there were serious or severe rainfall deficiencies over southern and eastern Australia, in an arc extending across southeastern South Australia, southwest, south-central and northeast Victoria, and the tablelands and western slopes in southeastern New South Wales. A large part of southeast Queensland was also affected, as were northern and eastern Tasmania, and Western Australia west of a line from Dampier in the north to Bremer Bay in the south (Figure 40).

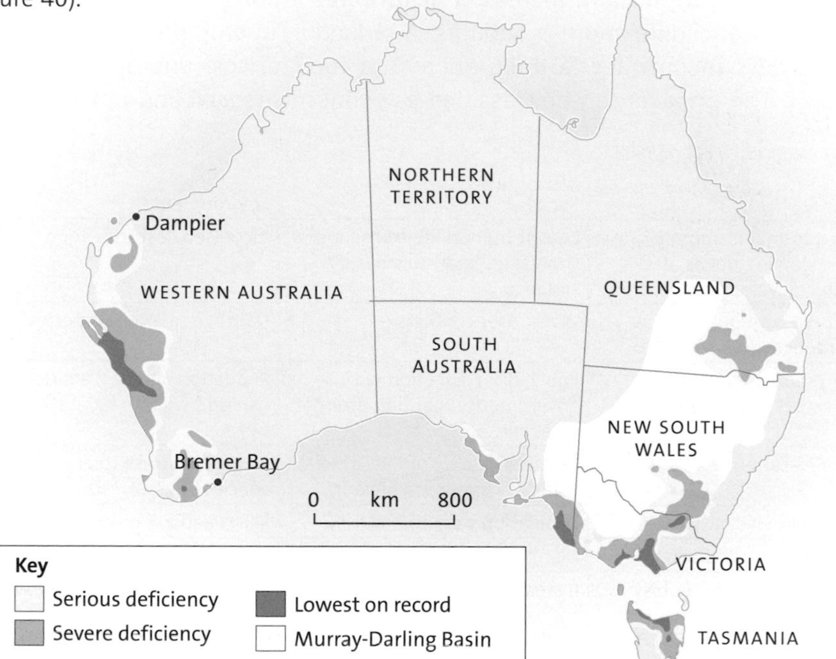

Figure 40 *Rainfall deficiencies in Australia, July 2006–June 2007*

The drought is affecting Australia's political environment as much as its agricultural landscape. The government has announced assistance for farmers of Aus$1.1 million, or £450 000, bringing the total over the past 4 years to almost 2.3 billion dollars (£1 million). This is mainly short-term assistance, in part to stave off farm repossessions by the banks, which many fear could spark a collapse in land prices.

The Australian government has come under considerable pressure, both at home and abroad, for its refusal to sign the Kyoto agreement (2007) on limiting greenhouse-gas emissions. This event has put the issue of drought and water supply firmly in the political arena. It has also caused rural–urban conflicts. For example, in the Murray–Murrumbidgee basin, the citizens of Adelaide were convinced that the fruit and rice farmers were taking too much water for irrigation.

THE NIGER FAMINE 2005: TRIGGERED BY DROUGHT?

Drought can be a trigger for famine especially when combined with biohazards, such as locust swarms (see *Case study 18*, p. 68). The Niger famine was one of the worst disasters of 2005 because it was so clearly predicted, 9 months ahead of the main tragedy.

The main causes of the famine can be summarised as:

- a premature end to the 2004 rains, resulting in drought
- damage to land by locusts, causing a lower crop yield than expected
- chronic poverty (with no safety net)
- high food prices resulting from food shortages

Arguably, the physical consequence of drought led to the famine hazard, but the warning signs were obvious. Only 45% of households in Niger's seven rural regions are food secure. Repeated warnings of the onset of famine were ignored, and demands for cash to pay for food aid and other relief were not met. Approximately 10 000 people died as a result of this drought-related disaster.

Managing the drought hazard

Modify the event

Drought cannot be prevented, although scientists have been experimenting with cloud seeding using silver iodide pellets to bring on rain storms. Dealing with drought mainly involves water storage and community preparedness, which includes attempting to predict the hazard.

Modify vulnerability

Droughts occur erratically, which makes planning difficult, but populations of dry lands have learnt to live with drought through careful management of resources.

Scientists use satellite imagery to measure the progress of rains in order to predict drought areas before people begin to starve due to crop failure. This method gives accurate warnings, but often aid agencies lack the resources to respond in time. Longer-term 'drought-aid' in the form of irrigation schemes and education for farmers on water conservation techniques is more useful.

Other strategies to alleviate drought include the transport and storage of food so that it can quickly reach affected populations.

In MEDCs, people are encouraged to adopt water conservation strategies, or governments may legislate to enforce such policies (for example, during the drought in southeast England in 2006). Richer societies may be able to afford large engineering projects such as dams and storage reservoirs, as well as desalination and water recycling plants.

Modify the loss

Land management schemes can help to reduce impact, although in LEDCs land degradation is tied up with social, political and economic factors. Dry farming techniques to conserve moisture are being used in some parts of the world. These involve better herd management and soil moisture conservation techniques.

Question

(a) 'A drought disaster is caused by the combination of both a climate hazard — the occurrence of deficits in rainfall and snowfall — and a societal vulnerability.' (World Bank, 2006) To what extent do you agree with this sentiment?

(b) Figure 41 shows the rainfall graph for Australia from 1900 to 2006. Describe the overall pattern of these data. How can these data be linked to drought?

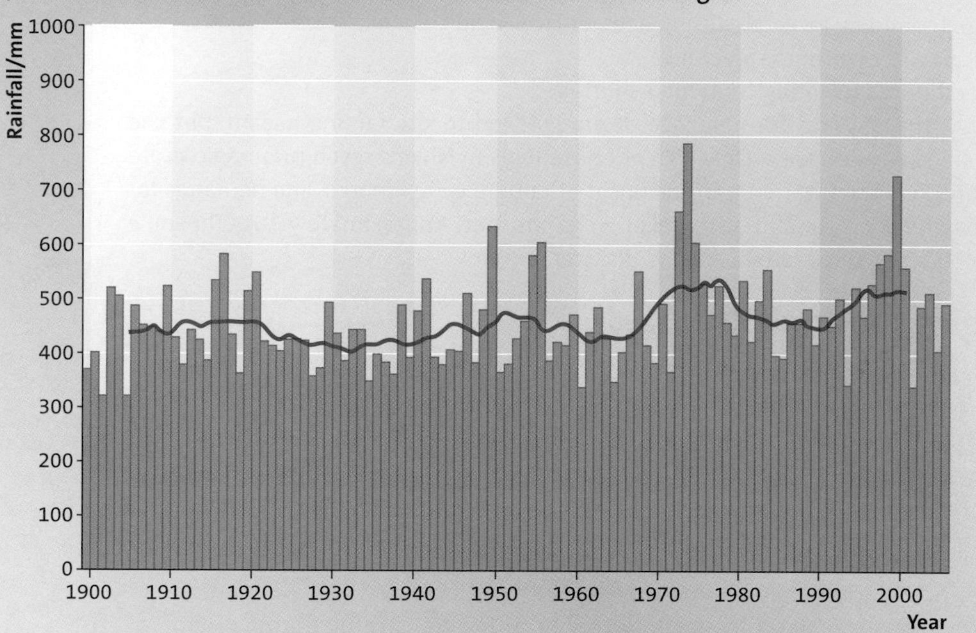

***Figure 41** Annual rainfall in Australia*

Guidance

(a) This is certainly true — the economic, social and political characteristics that render people and their livelihoods susceptible in any region influenced by the deficits or shortfall in rainfall.

(b) The graph shows the extreme variability of rainfall — from about 325 mm y^{-1} to 800 mm y^{-1}. With precipitation being so unreliable, droughts can occur, especially if successive years are comparatively dry and the hydrological system has little chance to recharge. The overall trend for rainfall may indicate a slight increase in precipitation. Drought might be more problematic due to rising temperatures leading to increased rates of evaporation and transpiration.

Hurricanes

The word 'hurricane' comes from the Spanish *huracan* and Carib *urican* meaning 'big wind'. These intense winds in the Atlantic and eastern Pacific Oceans are known by many different names. Only when the wind speed exceeds 120 km per hour (75 mph) does it become a true hurricane or **cyclone**.

Hurricanes and cyclones are predictable in terms of their spatial distribution (Figure 42). They are concentrated in the tropics, specifically 5° and 20° north and south of the equator. Once generated, they tend to move westwards and are at their most destructive in:

- the Caribbean/Gulf of Mexico and the western side of Central America — these areas have 28% of all cyclones (hurricanes)
- the Arabian Sea/Bay of Bengal area — 8% of all cyclones (tropical cyclones)
- southeast Asia and Madagascar — 43% of all cyclones (**typhoons**)
- northern Australia — 20% of all cyclones ('willy-willies')

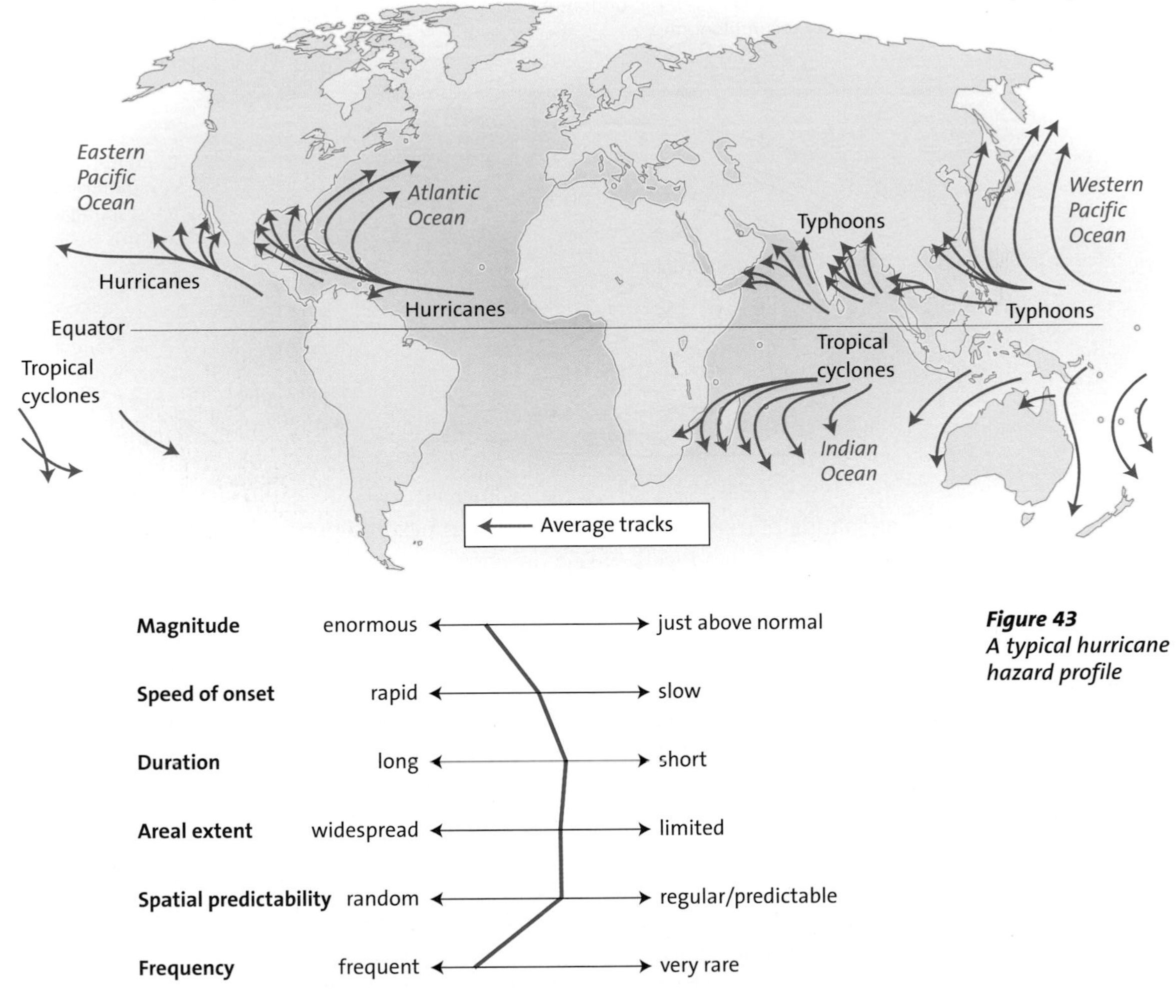

Figure 42 *Distribution of hurricanes and tropical cyclones*

Figure 43 *A typical hurricane hazard profile*

The causes of tropical cyclones are well known. They begin with an area of low pressure into which warm air is drawn in a spiralling manner. Small-scale disturbances enlarge into tropical storms with rotating wind systems, which may grow into a much more intense and rapidly rotating system the cyclone.

For a hurricane to develop:

- it must be in an oceanic location with sea temperatures over 26.5°C
- the location must be at least 5° north or south of the equator, so that the Coriolis effect can bring about the maximum rotation of air (the Coriolis 'spinning' effect is zero at the equator and increases towards higher latitudes)

- rapidly rising moist air (from the warm sea) cools and condenses, releasing latent heat energy which fuels the storm — hurricanes fade and die over land as this energy source is removed
- low-level convergence of air occurs in the lower circulation system — this, the inter-tropical convergence zone (ITCZ), is thought to be the precursor to tropical storms

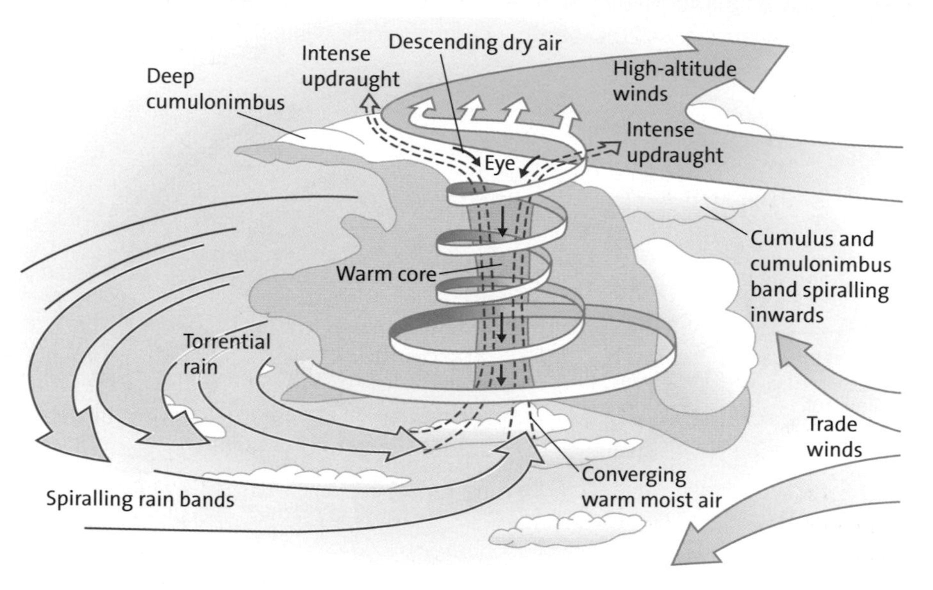

Figure 44
The development of a hurricane

Table 7
The Saffir–Simpson hurricane scale

The magnitude of tropical storms is typically measured on the **Saffir–Simpson** scale, which consists of five levels based on central pressure, wind speed, storm surge and damage potential. Table 7 shows the Saffir-Simpson scale.

Category	Wind speed (km per hour)	Effect	Storm surge (m above normal water levels)
1	119–153	No real damage to building structures; slight damage to trees and vegetation, some risk of coastal flooding	1.2–1.5
2	154–178	Some roofing material, door and window damage; considerable vegetation damage	1.8–2.4
3	179–209	Some structural damage to small houses and utility buildings; extensive coastal flooding	2.7–3.6
4	210–249	Extensive damage; complete roof collapse possible for small houses; extensive coastal erosion and flooding extending well inland	3.9–5.5
5	249+	Complete roof failure on many dwellings and industrial buildings; major flood damage; massive evacuation of residential areas may be required	5.5+

The impacts of hurricanes/cyclones include:

- winds exceeding 150 km per hour (up to a maximum of 250 km per hour) cause structural damage and collapse of buildings, damage to bridges and road infrastructure and loss of agricultural land
- heavy rainfall, often over 100 mm per day, causes severe flooding and sometimes landslides (see *Case study 16* on p. 59); high relief may exaggerate rainfall totals, causing them to exceed 700 mm in a single day

- storm surges result from the piling up of water by wind-driven waves and the ocean rising up under reduced pressure — flooding can extend inland if the area near the coast is flat and unprotected

The impact depends on the physical and human factors outlined in Table 8.

***Table 8** Factors influencing impacts of hurricanes*

Physical	Human
Intensity of the cyclone (magnitude on the Saffir–Simpson scale)	Population density of the affected area
Speed of movement, i.e. length of time over area — slow movement = flooding	Level of community preparedness and education
Distance from the sea as hurricanes decline because of friction on land	Ability to predict and warn about the impending hazard; quality of forecasting
Physical geography of coastal impact zone, i.e. location of mountains near coast versus flat deltas	Nature and construction of buildings; effectiveness of local, regional and national government

TYPHOON NITANG/IKE

Case study 10

Typhoon Nitang/Ike was responsible for one of the worst natural disasters in the Philippines in modern times. The storm struck the east coast with 240 km per hour winds on 1 September 1984 and rapidly cut a swathe of destruction across the archipelago.

Twenty-three provinces in seven regions were affected. The southern provinces of Surigao del Norte, Bohol Island and Negros Occidental and the central/southern Cebu provinces were worst hit. An estimated 547 067 people were made homeless by floods and strong winds. Heavy rainfall and winds killed 1492 people and more than 1.2 million were affected. The damage costs (equivalent to 2007) were in the region of US$145 million or £70 million.

***Figure 45** The track of typhoon Nitang/Ike uses the colour scheme from the Saffir–Simpson scale (see Table 7)*

STORMY TIMES DOWN UNDER

Case study 11

On 20 March 2006, Australia's worst tropical storm since records began made landfall on the Queensland coast, close to the town of Innisfail. Gusting at up to 290 km per hour, Cyclone Larry was a category 4–5 storm on Australia's tropical cyclone severity scale.

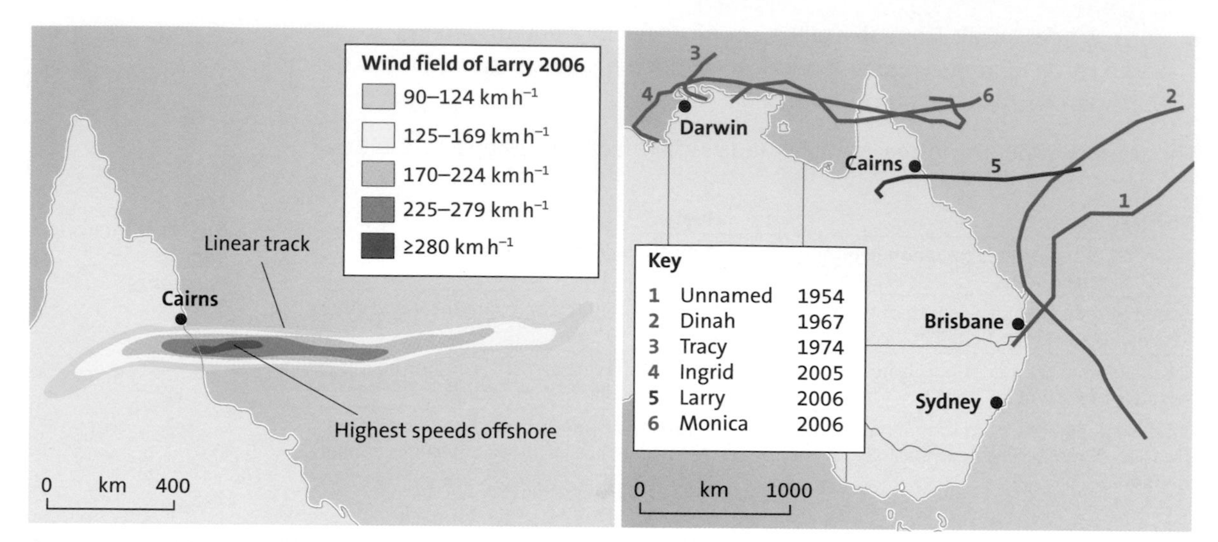

Figure 46 *Wind field of Cyclone Larry*

Figure 47 *Tracks of tropical cyclones in Australia*

More than 10 000 buildings in the cyclone's path were damaged or destroyed and 120 000 homes were hit by power cuts. Total losses were estimated at £500–600 million, including destruction of banana and sugar-cane plantations in the order of £200 million. The cyclone destroyed a large proportion of Australia's banana crop. Australia is relatively free of banana pests and diseases, and therefore does not allow bananas to be imported. Bananas were in short supply and expensive throughout Australia for the remainder of 2006, with prices rising as high as Aus$13–14 (approximately £4–£5 in 2006) per kilogram.

Cyclone Larry was an unusual event. Australia had not faced a direct hit on a city since the 1970s. Some scientists fear that global warming will increase the frequency and magnitude of cyclones, and also cause them to occur outside the main locations as seas are warmed by the enhanced greenhouse effect. Figure 47 shows the tracks of recent tropical cyclones in Australia.

Managing the tropical cyclone hazard

Modify the event

There is ongoing research into how tropical storms can be tamed. Much of this effort is directed at ways of reducing the storm's energy while it is still over the ocean.

Project Stormfury was an attempt to weaken hurricanes. Silver iodide was used to 'seed' the storm outside the eye-wall clouds. The idea was to produce rainfall, so releasing latent heat that would otherwise sustain the high wind speeds. Instead, a new eye-wall was created further out to sea, with lower wind speeds. The project has had limited success as 'seeded' storms have proved difficult to manage.

Modify vulnerability

The prediction of tropical storms depends on monitoring and warning systems. It is essential that warnings are correct because of the high economic cost associated with evacuation. The USA maintains round-the-clock surveillance of tropical storms using weather aircraft.

While monitoring occurs and predictions can be made in LDCs, it is not always possible to give more than 12–18 hours' warning because communications are poor. This is insufficient for a proper evacuation.

Some areas do have established warning systems, which give people the chance to take precautions. Examples include Belize in Central America and the Bay of Bengal in India and Bangladesh.

Forecasting the precise power and track of a hurricane remains problematical, so storm warnings are not always accurate or reliable. As with many hurricanes, it is not only the primary impact of the storm, but also the secondary hazard of storm surges and widespread coastal flooding that lead to extensive damage and loss of life.

In the case of Hurricane Katrina (2005), widespread flooding was made worse by inadequate flood defences. Some experts believe that Katrina's impact would have been reduced if the marshes that once protected the Louisiana coastline had still been there to shield it from the storm. The marshes have been stripped away by development, leaving the city of New Orleans vulnerable. The Mississippi River, which drains 40–50% of all the water from continental USA, has been prevented from depositing its natural silt load near the coast. Dredging channels to reach oil and gas wells has made matters worse. Post Katrina, pilot schemes to restore small areas of wetland are encouraging the development of native plants. It is hoped that the wetlands will act as natural buffers and reduce the energy and impacts of future hurricanes.

Figure 48 *Hurricane forecast errors*

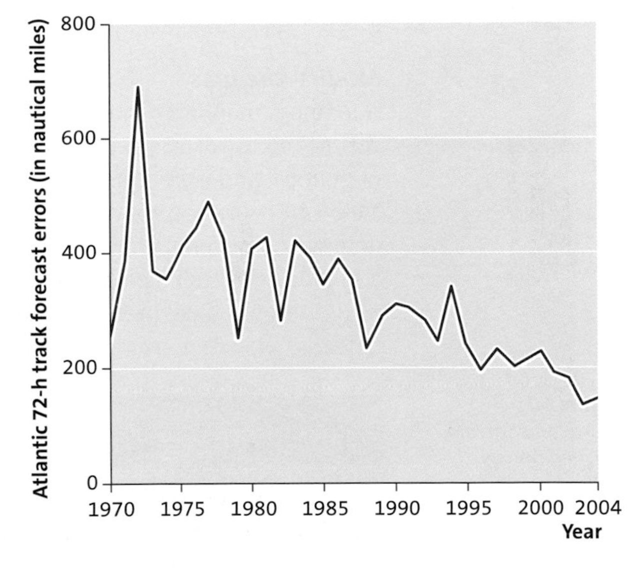

Better forecasting of cyclone tracks and intensities (Figure 48) could reduce deaths and property damage by enabling the issue of more timely and accurate warnings and evacuation orders. This is an important aim for the USA: in the memorable 2005 hurricane season, more than 1000 people died and the costs totalled more than $100 billion.

Scientists are also aiming to build better computer forecasting models. New technologies allow forecasters to break storms into a grid, and use sophisticated methods to predict changes in wind speed, humidity, temperature and cloud cover. Current NOAA models are too inaccurate to predict where the greatest damages might occur. The high-resolution NOAA Hurricane Research and Forecasting model became operational in 2007 (Figure 49).

Preparedness is the best form of protection. Accurate predictions enable evacuation to take place and emergency services can be put on full alert. Schoolchildren in Florida practise hurricane drills, similar to those carried out in earthquake-prone areas, as part of a hurricane awareness programme called Project Safeside.

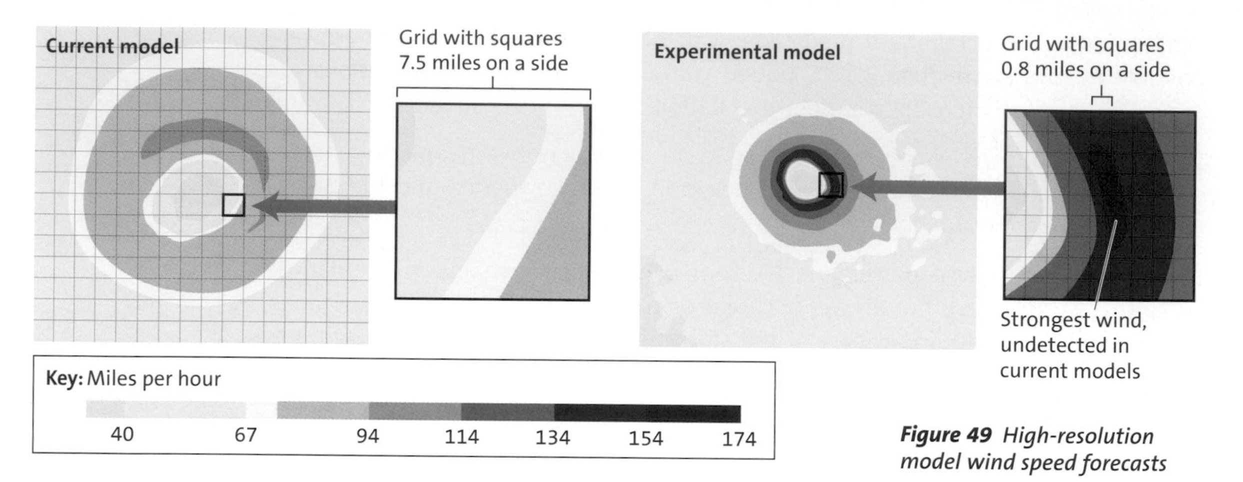

***Figure 49** High-resolution model wind speed forecasts*

Other protection strategies include:

- land-use planning, so that areas of highest risk have limited development
- strengthening of buildings to withstand windstorms and floods. Buildings in Dominica in the West Indies were retrofitted to make them resistant to winds in 1994 and successfully withstood the impact of Hurricane Marilyn the following year.
- sea walls, breakwaters and flood barriers, and houses built on stilts

Modify the loss

Adequate insurance before the disaster, and aid during and after the event, contribute towards modifying the loss. The impact of cyclones depends on a range of political and economic factors. Areas with lower levels of development suffer from lack of insurance, poor land-use planning, inadequate warning systems and defences, and poor infrastructure and emergency services. This usually results in a higher death toll (see Table 5 on p. 26). Hurricane Katrina exposed the problems of the largely uninsured, relatively poor population within the MEDC city of New Orleans, who were trying to cope during and especially after the event. The problem

***Figure 50** Hurricane Katrina pushed houses inland on the Mississippi coast, for example at Biloxi*

of organising aid effectively, which is vital to overcome the loss of property and livelihoods, is discussed in *Case study 26* (p. 94).

Impact of climate change on tropical cyclones

Controversy rages over whether global warming is making hurricanes stronger. An increase in the incidence and intensity of hurricanes was previously attributed to the natural longer-term climatic cycle. Recently, however, scientists have begun to think that a surge in sea temperatures since 1970 has made hurricanes more severe. In particular, hurricanes:

- are more intense (of greater magnitude)
- last longer
- yield more precipitation
- have a less predictable pathway
- *but* have a reduced or similar frequency

Figure 51 ***Hurricane intensity and sea temperature variation (combined figures for the North Atlantic and western North Pacific)***

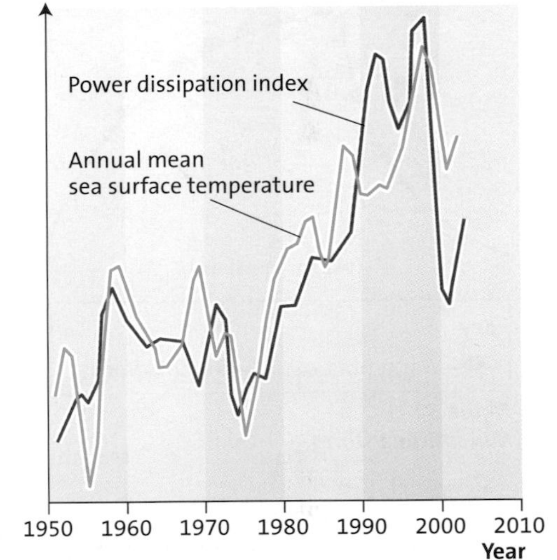

Figure 52 suggests a strong correlation between higher sea surface temperatures and more intense storms, but sea temperatures change due to the El Niño–La Niña cycle, which has a major impact also on magnitude and frequency of hurricanes.

Other scientists argue that it is impossible to know whether recent hurricanes have been made worse by global warming — a much longer future record is required. But many agree that there is a 'human fingerprint' (i.e. influence) in hurricane trends. This might change not only the nature of hurricanes but also their distribution, especially into areas with cooler waters that were not previously considered to be spawning grounds. There was therefore great concern when Hurricane Catarina hit the coast of southeast Brazil in 2004.

Using case studies 8

Question

(a) To what extent can tropical cyclones be prevented?
(b) Read the section 'Impact of climate change on tropical cyclones' above. How strong is the link between climate change and hurricane intensity and frequency? How might climate change influence hurricanes in the future?

Guidance

(a) Tropical cyclones cannot be prevented — the best that can be hoped for is modification in terms of severity/intensity and altering the path a storm may take to reduce its effects (e.g. diversion back out to sea). There is concern that tampering with this complex system might cause unknown effects on the Earth's global energy system. There is also the question of liability — for example, diversion of a storm into a neighbouring state or area. However, modification of impacts carries large potential rewards that should not be overlooked.

(b) There is a probable link between intensity and increasing sea temperatures, but not necessarily frequency. The science is complex and disputed. The spatial extent of cyclone development may increase, and there may be more 'unusual' events.

Extra-tropical storms

Temperate, mid-latitude or extra-tropical storms affect large areas of the world outside the belt where tropical storms are usually found. They become a significant hazard only when they affect densely populated areas such as parts of Europe.

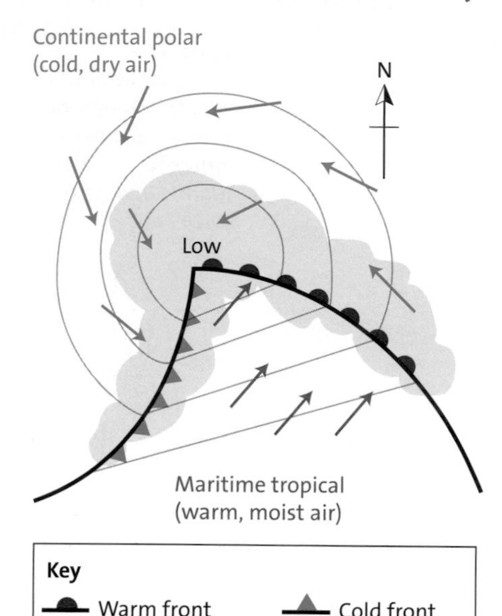

Figure 52
Mid-latitude storm

Mid-latitude storms can have a range of impacts:

- strong winds, often over 100 km per hour
- heavy rainfall — especially when combined with storm surge conditions
- coastal flooding under low pressure and strong onshore winds
- in winter, heavy snowfalls replacing rain
- sometimes the development of mini-tornadoes (see p. 50)

Mid-latitude storms can cause widespread damage. Several storms may cross Europe in quick succession. This can lead to secondary hazards, such as flooding, when catchments are already saturated. Depressions can track over highly populated areas and have the potential to cause economic damage. They may disrupt power supplies and telecommunications. There may be impacts on agriculture, which can affect the economy, for example food prices. Mid-latitude storms can be spatially large, sometimes affecting much larger areas than hurricanes.

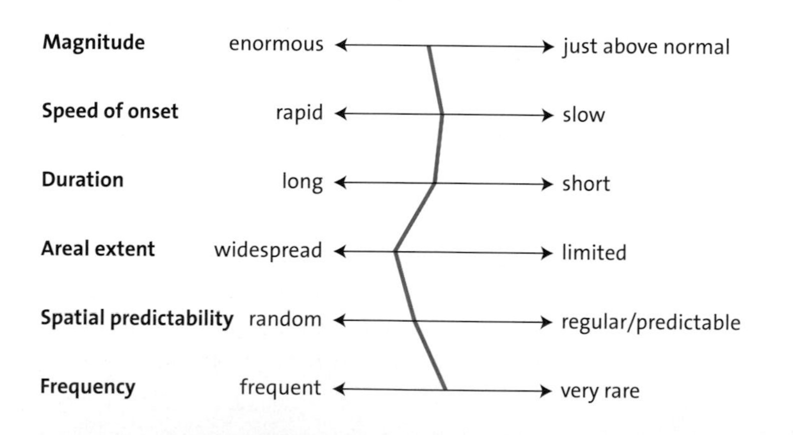

Figure 53
A typical extra-tropical storm hazard profile

THE 'GREAT STORM' OF 1987

The worst UK storm in nearly 300 years hit the southern part of England in October 1987. Figure 54 shows the synoptic chart for that time. At the storm's peak, the wind speed consistently exceeded 80 km/h, leaving a trail of destruction in its wake. Conditions at sea were even more alarming: the Royal Sovereign lighthouse on the south coast recorded a mean wind speed of nearly 140 km per hour. Gusts of 150 km per hour occurred around the coasts and 140 km per hour was exceeded inland.

An estimated 15 million trees were blown down and thousands of homes were without electricity for more than 24 hours. Buildings and vehicles were crushed by falling trees. Numerous small boats at sea were wrecked or blown away, a ship capsized in Dover harbour, and a Channel ferry was driven ashore near Folkestone. Eighteen people died but the toll would have been much worse had the storm arrived during daylight hours. The total cost of the damage ran into billions of pounds.

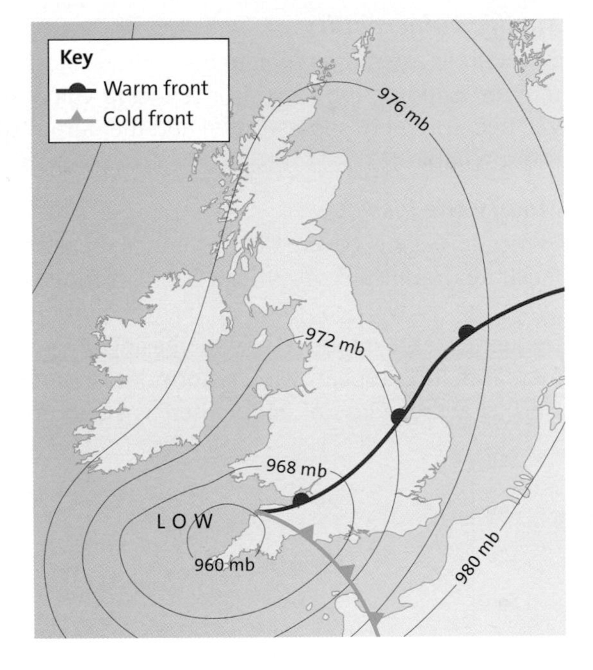

Figure 54
Synoptic chart for the UK at midnight, 16 October 1987

'STORM OF THE CENTURY': EASTERN USA

Case study 13

In early March 1993, weather satellite photographs showed a large mass of cold air moving down from the North Pole across North America. This mass of cold air eventually collided with warmer air in the region above the Gulf of Mexico.

A line of powerful thunderstorms formed along the front, drawing energy from the temperature differentials. The winter storm moved onto land during the early hours of Friday morning, 12 March, killing dozens of people and devastating parts of the Florida coast.

The storm then began to travel up the east coast. As it crossed the eastern seaboard, torrential rain turned to heavy snow, falling from Alabama to New York and virtually paralysing the eastern third of the country. Local authorities were totally unprepared for the intensity of the storm. The interstate highways became impassable and millions of people lost electrical power. New York City was brought to a standstill. A foot of snow fell from Alabama to Maine, and freezing temperatures set new records across the eastern seaboard.

The final account included 243 deaths, and about US$2 billion in damage. The storm had forced the closure of all the airports in the eastern USA, and created chaos. Nearly 100 million people in 26 states were affected.

Managing the mid-latitude storm hazard

Modify the event

The normal responses include avoiding the hazard, modifying the causes, reducing the effects or doing nothing. In the case of extra-tropical storms, little can be done to modify these gigantic weather systems.

Modify vulnerability

Some areas are more prone to mid-latitude storms than others. Prediction is a useful tool for avoiding the hazard or reducing vulnerability. In future, computer models will become more sophisticated and the output will probably be more detailed and more reliable.

Modify the loss

Storm damage is covered by insurance. Insurance provides economic security and facilitates rapid recovery by offering compensation. However, the global insurer Swiss Re is worried that, in the medium to longer term, mid-latitude storm frequency and intensity are increasing. Figure 55 shows the estimated annual increases in European winter storm losses between 1975 and 1985. Note that the UK has average loss of 25%, whereas Denmark has nearly 120%.

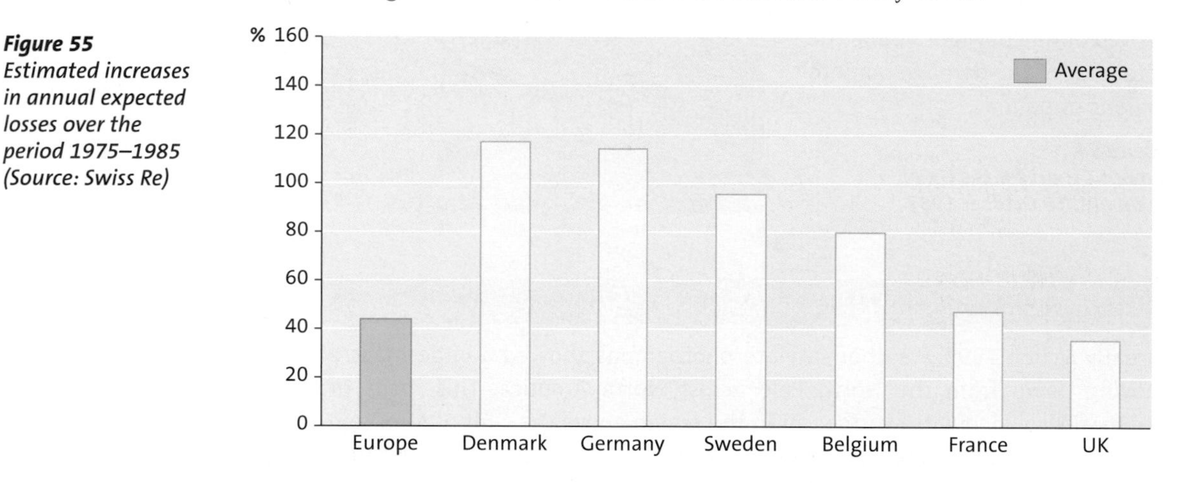

Figure 55 *Estimated increases in annual expected losses over the period 1975–1985 (Source: Swiss Re)*

9 Using case studies

Question

(a) How are tropical cyclones different from mid-latitude storms?
(b) What impact is climate change likely to have on mid-latitude storms?

Guidance

(a) Tropical cyclones:

- are low-pressure systems that derive energy primarily from evaporation from the sea in the presence of high winds and lowered surface pressure. There is associated condensation in convective clouds concentrated near their centre.
- have their strongest winds near the Earth's surface (a consequence of being 'warm-core' in the troposphere). 'Warm-core' refers to being relatively warmer than the environment at the same pressure surface ('pressure surfaces' relate to height or altitude).

Mid-latitude storms:

- are low-pressure systems with associated cold fronts, warm fronts and occluded fronts
- derive their energy primarily from the horizontal temperature gradients that exist in the atmosphere
- have their strongest winds near the tropopause (a consequence of being 'warm-core' in the stratosphere and 'cold-core' in the troposphere)

There are other differences in terms of scales of impacts, frequency etc.

(b) Statements often appear in the media suggesting that more extreme mid-latitude storms will result from global warming. However, climate models are limited in their ability to represent mid-latitude storms adequately. There is some evidence for global warming leading to 'polar amplification' (stronger warming near the North Pole), and so the average poleward temperature gradient is expected to diminish, leading to more stable conditions on average. Increased temperatures also imply higher humidity, and thus a higher capacity for energy conversion through condensation — the energy fuel of convection — contributing to more storms. One certainty is that insurance premiums will rise.

Tornadoes

A tornado is the most violent of atmospheric storms. It consists of a rotating column of air extending downwards from a cumulonimbus (thunderstorm) cloud. Tornadoes are small and short lived, but they have very high energy and are capable of doing considerable damage.

The centre of a tornado is characterised by extremely low pressure, which sucks up dust to form a dark-grey funnel rising into the sky. Around this funnel of rising air are strong winds capable of destroying crops and buildings.

Tornadoes result from the interaction of contrasting warm and cold air masses over land. They tend to be restricted to mid-latitude areas, classically the central areas of the USA (Figure 57) where warm air from the Gulf of Mexico meets colder Arctic air from the north.

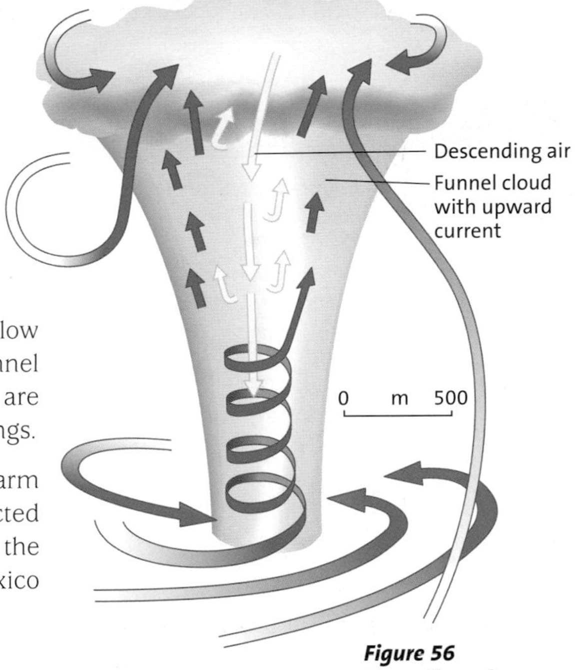

Figure 56 *Formation of a tornado*

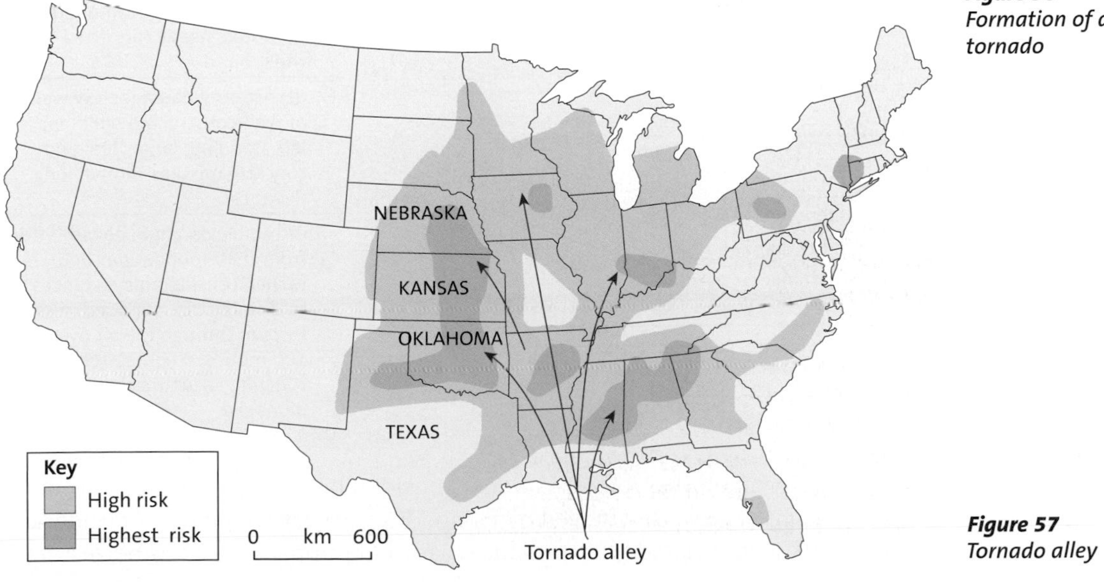

Figure 57 *Tornado alley*

newsteam.co.uk

Figure 58 *A mini tornado sweeps across fields in Shropshire during extreme weather in June 2007*

Although 80% of recorded tornadoes occur in the USA, other regions of the world are not immune (see Figure 58). They tend to be less common over towns and cities because the frictional effect of buildings reduces the likelihood of their development. They can also occur at sea as water spouts.

The main impacts associated with tornadoes are due to the high wind speeds, which can reach over 500 km per hour (139 m s^{-1}). The **Fujita** scale is used to describe the nature of damage associated with various strength tornadoes (Figure 59). This allocates a tornado to one of five categories, according to average wind speed.

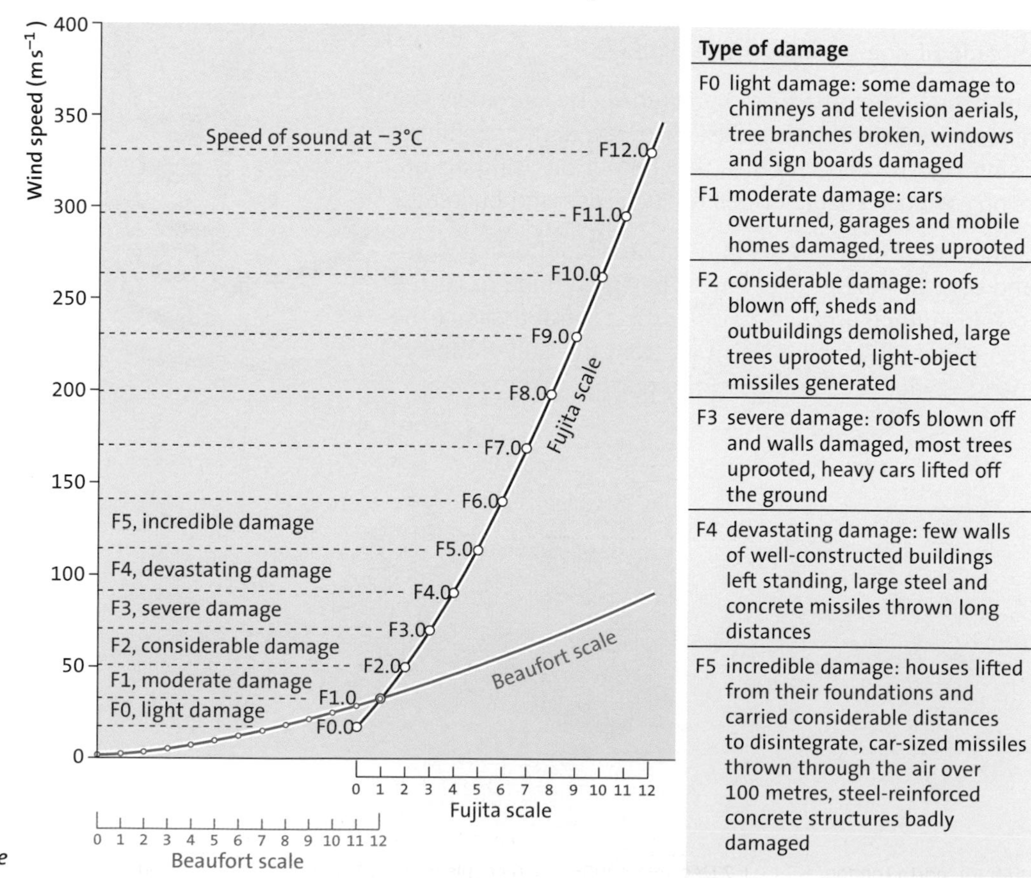

Type of damage	
F0	light damage: some damage to chimneys and television aerials, tree branches broken, windows and sign boards damaged
F1	moderate damage: cars overturned, garages and mobile homes damaged, trees uprooted
F2	considerable damage: roofs blown off, sheds and outbuildings demolished, large trees uprooted, light-object missiles generated
F3	severe damage: roofs blown off and walls damaged, most trees uprooted, heavy cars lifted off the ground
F4	devastating damage: few walls of well-constructed buildings left standing, large steel and concrete missiles thrown long distances
F5	incredible damage: houses lifted from their foundations and carried considerable distances to disintegrate, car-sized missiles thrown through the air over 100 metres, steel-reinforced concrete structures badly damaged

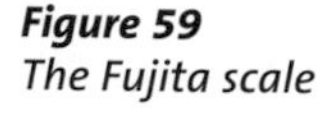
Figure 59
The Fujita scale

The lifting force of the funnel can be considerable. Large vehicles, including railway carriages, are easily moved by F4 and F5 category tornadoes. Considerable flying debris is associated with tornadoes. Tornado impact tends to be restricted

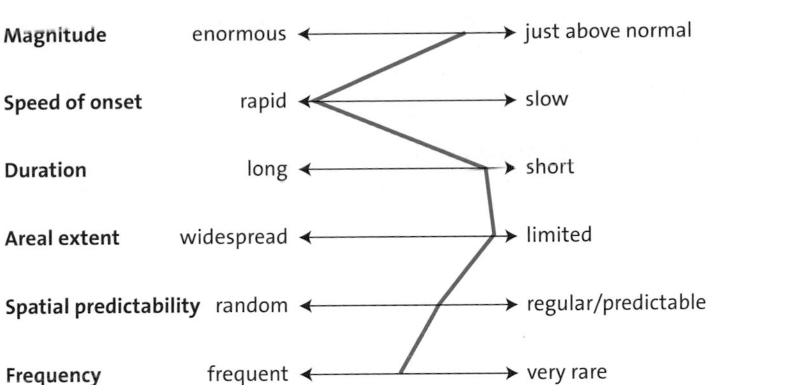

Figure 60
A typical tornado hazard profile

to a small area. Buildings just outside this narrow zone might escape with minor structural damage.

A number of different types of 'whirlwind' are reported in the UK every year. The year 2005 was particularly active, with 208 events recorded.

TORNADO IN BIRMINGHAM, UK

Case study **14**

In July 2005, 19 people were injured as a tornado tore through the Kings Heath area of Birmingham. The storm damaged buildings, uprooted trees and trapped people in their homes. The Met Office estimated that the wind speed reached 209 km per hour. The tornado was estimated to be F2–F3 (with wind speeds of nearly 180 km per hour). The damage was concentrated in the Balti belt area. Many small Asian businesses took up to 2 years to recover.

This event was unusual because tornadoes rarely form over built-up areas.

Table 9
Some notable UK windstorms

Location	Date	Notes
Rosdalla, Ireland	30 April 1054	Earliest known European tornado
London, England	23 October 1091	Earliest recorded British tornado
Fernhill Heath, Hampshire, England	22 September 1810	One of the most well-known and strongest British tornadoes
Plymouth, England	14 December 1810	Reputed to be the strongest tornado of all time in southwest England
Tay Estuary, Scotland	28 December 1879	79 fatalities; a water spout destroyed the central section of Tay Bridge, killing everyone on board a passing train
Barry and Chester, Wales and England	27 October 1913	A tornadic supercell with deadly consequences; several fatalities
Buckinghamshire and Cambridgeshire, England	21 May 1950	A long-lasting tornado (up to 3 hours) with several 'branches'
Gwynedd, Humberside and Essex, Wales and England	21 November 1981	Largest known European outbreak; 105 predominantly weak tornadoes were recorded
Birmingham, England (see *Case study 14*)	28 July 2005	Wind speeds of up to 180 km per hour; 19 people injured and hundreds of properties affected
Various locations in England	August 2006	More than 4 tornadoes recorded; 8 people injured
Kilburn, north London, England	7 December 2006	6 people injured, 100 properties damaged
Various locations in England	15 June 2007	Severe storms brought heavy flooding to the Midlands and Yorkshire; at least 10 tornadoes across England, possibly including a wide wedge tornado in Cornwall

Scientists believe tornadoes will become more common in the UK and in such a heavily populated island, urban areas will be inevitably hit. This will lead to high insurance costs.

Managing the tornado hazard

Modify the event

There is little that can be done to prevent tornadoes occurring, although some researchers are suggesting that modifications in land use may provide less opportunity for tornado development. On the Canadian prairies, for example, a significant portion of the transpiration is derived from agriculture, which is dominated by spring wheat and other field crops with similar water use patterns. By changing the humidity of the atmosphere through a modification of crop type, the potential energy available for deep convection development can be altered. This has an impact on the likelihood of tornado development.

Modify vulnerability

Prediction is vital with such violent and sudden events (tornadoes can form very quickly, emerging in just 5–10 minutes), but it may never be possible to predict the precise location and timing of a tornado, even just 24 hours into the future. This means that little warning can be given to evacuate.

General warnings can be given using information from radar and satellite technology. The USA has formal tornado forecasting. Warnings are issued on the internet, television and radio.

In regions where tornadoes present a frequent threat, individuals, businesses and other organisations may have their own tornado safety plans. This might involve evacuation to less vulnerable parts of buildings, or to a purpose-built safe area or building. Some regions have 'mass care shelters', which can hold hundreds of families.

Modify the loss

Insurance to cover the cost of tornado damage is an important part of wider extreme weather protection. Scientists are working with computer programmers to help the insurance industry calculate premiums and risks. Computer simulation technologies are used to estimate the probability of damage from tornado events. With some major tornadoes, such as the one that caused a sea of devastation and destroyed the town of Greenburg, USA in 2007, federal and state programmes are vital immediately after the event. These are especially important for the poor who are uninsurable or who are under-insured.

Increasing global temperatures may provide the ingredients for more thunderstorm activity and hence more tornadoes. As land areas become more crowded with human settlement, the chance of a tornado strike is likely to increase.

10 Using case studies

Question

(a) Why is the number of deaths from tornadoes relatively small?

(b) What impact might climate change have on the frequency, intensity and distribution of tornadoes?

Guidance

(a) The path of a tornado is typically short and narrow, affecting a limited number of properties. However, 308 people died in the USA from the impacts of the deadliest tornado ever recorded, in April 1974.

In most parts of the USA, good prediction and warning systems exist to reduce the vulnerability of the hazard.

(b) Research in the USA shows a general increase in the number of tornadoes between 1950 and 1999 (Figure 61). This might be due to improved reporting, but it could be due to changing climatic conditions. There may be a link between El Niño/La Niña and changing patterns of frequency and intensity, but there is not enough statistical evidence to confirm this.

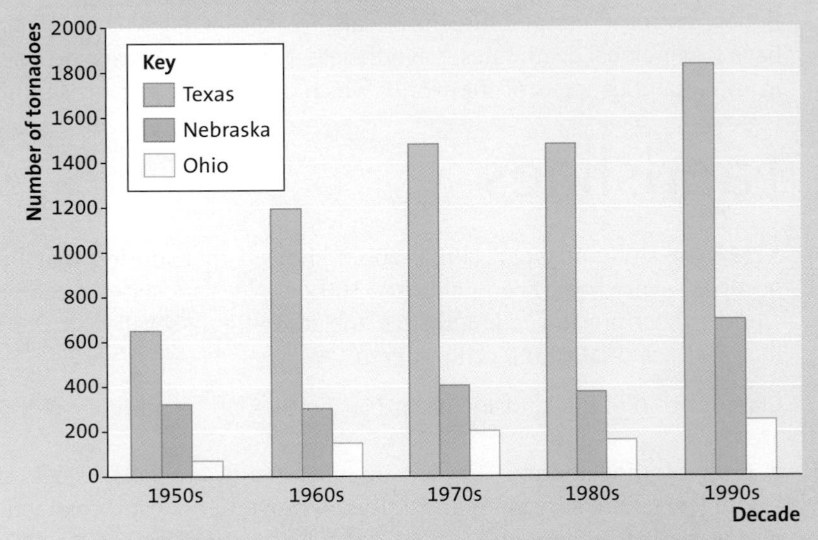

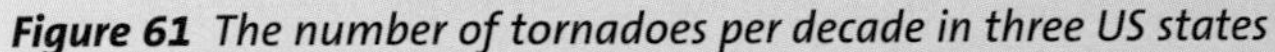

Figure 61 *The number of tornadoes per decade in three US states*

Part 4

Geomorphic hazards

Landslides and avalanches are the seventh biggest killer worldwide, ranking below virtually all other natural hazards (including windstorms/cyclones, wave surges, floods, extreme temperatures and earthquakes). Only volcanoes and forest fires have lower global death rates. Nevertheless, they are a widespread hazard, especially in mountainous areas of the world, which are subject to heavy rainfall.

Landslides

A 35-year study of Japan (1967–2002) showed that, during that time, landslides occurred every year, killing almost 3300 people. Landslides threaten some of the world's most precious cultural sites, including Egypt's Valley of the Kings and the Inca fortress of Machu Picchu in Peru.

Classification of the landside hazard is complex and a number of different systems are used:

- **Type of movement** — the main movements are falls, slides and flows, but topples, lateral spreading and complex movements can be added to these.
- **Material of movement** — rock, earth and debris are the terms used to distinguish the materials involved in landslides. The distinction between earth and debris is made by comparing the coarse grain size fractions.
- **Activity** — the concept of activity is linked to state, distribution and style. 'State' describes information regarding the time in which the movement took place, 'distribution' describes where the landslide is moving, and 'style' indicates how it is moving.
- **Velocity of movement** — a velocity range is linked to different types of landslide.

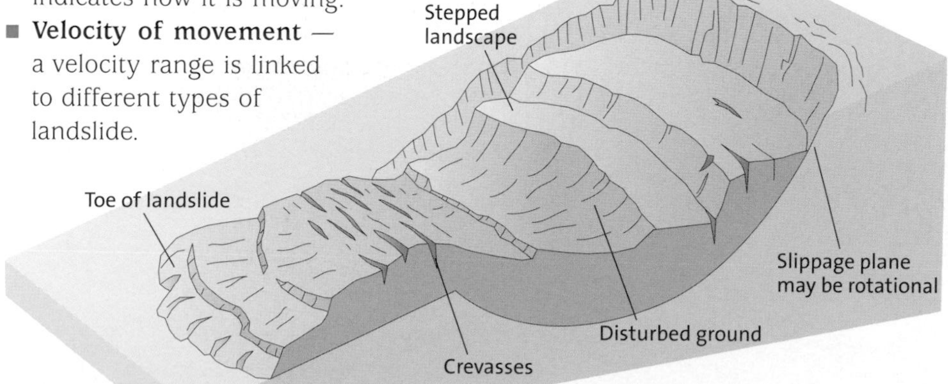

Figure 62
The landslide hazard (usually a small-scale feature)

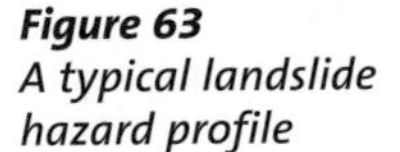

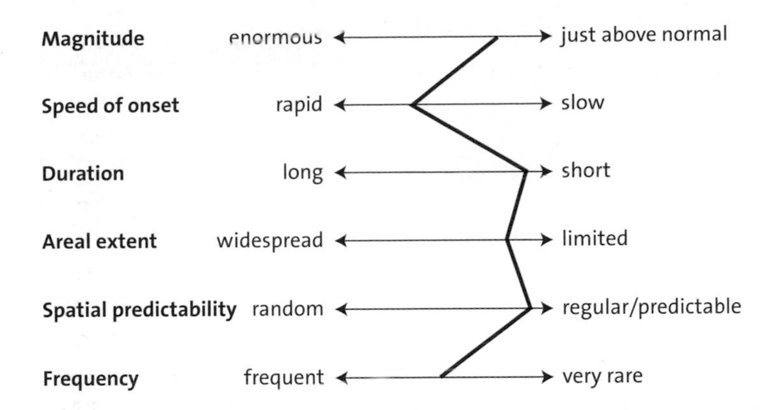

Figure 63
A typical landslide hazard profile

Landslides are caused by large volume combinations of rock, soil, ice, water and snow moving under the influence of gravity and, usually, water. Landslides are commonly composed of rock and soil, whereas the term avalanche refers to falls made up mainly of snow and ice (see p. 56).

Table 10 lists some of the worst landslides ranked by deaths.

Location	Year	Cause	Estimated death toll
Kansu, China	1920	Earthquake	180 000
Kahait, Tajikistan	1949	Earthquake	12 000
China	1998	Heavy rain	3 600
Chiavenna Valley, Italy	1618	Heavy rain	2 240
Nicaragua	1998	Heavy rain	2 200
Pakistan	1998	Heavy rain	1 500
Philippines	2006	Heavy rain/earthquake	1 100
China	2000	Heavy rain	625
Rio de Janeiro, Brazil	1966	Heavy rain	550
Goldau Valley, Switzerland	1806	Ice peak broke off	500

Table 10
Ten deadly landslides

Landslide distribution is controlled by a number of localised physical factors, including gradient and topography, land use, geology, permeability and climate. There are also some human-related factors that can promote the onset of landslides (Figure 64).

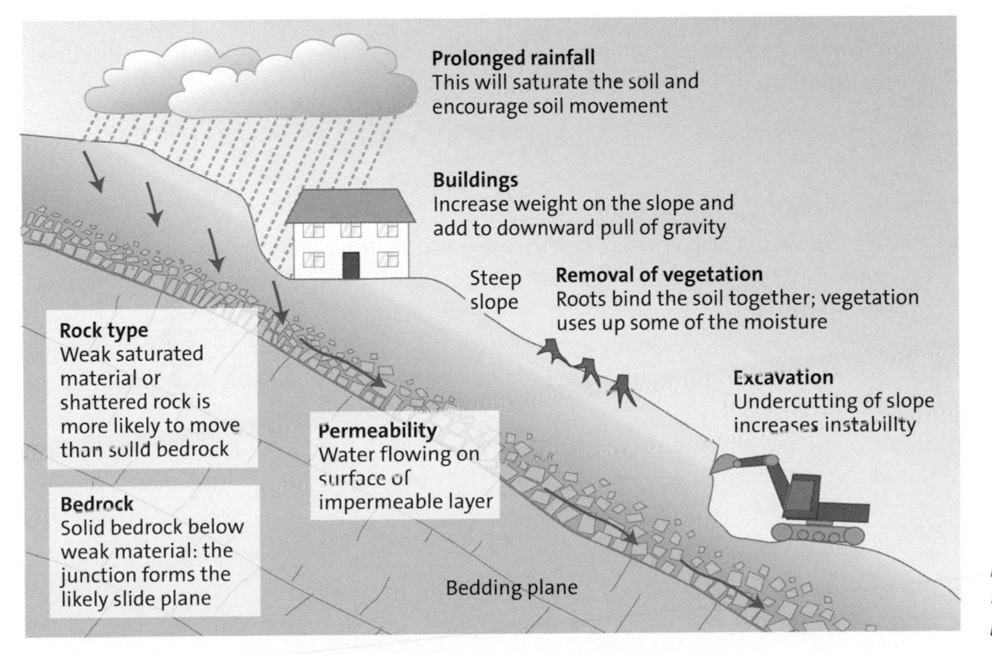

Figure 64
Factors that promote landslides

Table 11
The main causes of landslides in the UK

Process/causal feature	Recorded number
Weathering — loose material (regolith)	647
Natural erosion, e.g. by river or sea at base	1297
Artificial erosion, e.g. a railway cutting	276
Ground subsidence and removal of support	103
Deposition, e.g. of colliery waste on hillside	126
Seismic shocks and vibrations	73
Human-induced water regime change leading to surplus water lubricating soil	1203

Case study 15 CHITTAGONG MUD SLIDES

Bangladesh's 2007 monsoon started with unusually heavy rain, intensified by a storm from the Bay of Bengal on 9–10 June. The heavy rain caused mudslides, which engulfed slums in the foothill areas of the large city of Chittagong on the coast in southeastern Bangladesh on 11 June 2007. The death toll was reported to be at least 128, including 59 children, and more than 150 people were injured.

Experts had warned of the increasing likelihood of landslides due to the Bangladesh government's failure to curb illegal hill cutting in Chittagong. Hill cutting creates flat sites for the construction of new houses. Trees have also been cleared to build houses on the hill tops. This can block the natural rills or gullies that drain the landscape. Rainwater is then forced to enter the ground through cracks, which weakens the soil structure and promotes landslides.

Avalanches

Avalanches are a common phenomenon throughout mountainous areas of the world. Their distribution is becoming more widespread as human activity in these areas — mainly recreation and leisure tourism — increases. Specialist infrastructure to support such tourism is also increasing.

Avalanche distribution is relatively predictable because avalanches tend to recur in the same places.

There are three types of avalanche hazard:

- Powder snow avalanches can occur with little or no warning, at any time in the season, with speeds up to 300 km per hour and a force of up to 50 tonnes per square metre.
- Wet snow avalanches usually occur late in the season and are slower moving (8–25 km per hour). They typically carry a considerable weight of snow — up to 1 million tonnes.
- Slab avalanches are the most commonly occurring type and are often started by human error. With speeds of up to 150 km per hour, they cause death among skiers and snow-boarders.

The key causes of avalanches are:

- the weather — snowfall is an essential ingredient
- slope — more than 30° and less than 45° for starting an avalanche

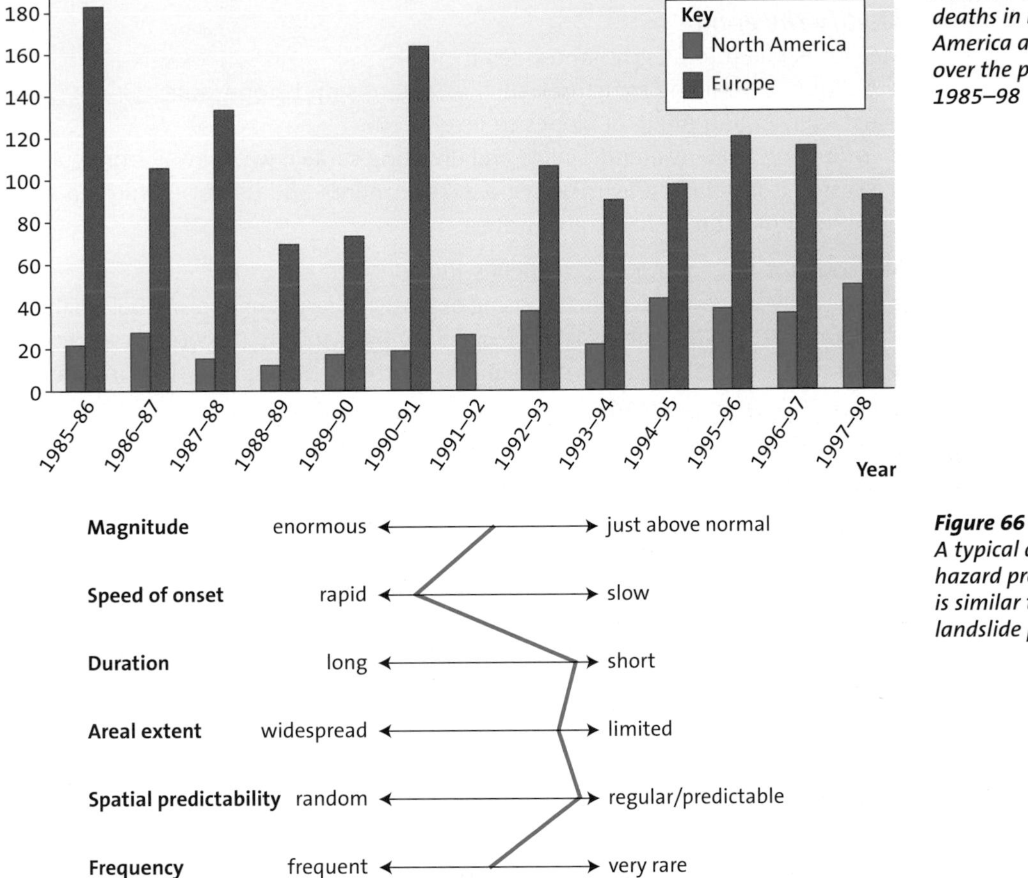

Figure 65 *Avalanche-related deaths in North America and Europe over the period 1985–98*

Figure 66 *A typical avalanche hazard profile, which is similar to the landslide profile*

- changes in the snow pack. Air temperature changes can bring about partial melting of the snow pack or changes in snow crystal shapes and sizes. This influences the strength of the layers. There is often a delicate balance between the tensional forces keeping the snow anchored to the slope (adhesion and weight) and the forces trying to promote movement

When a combination of factors act together, rapid mass movements can occur.

For both landslides and avalanches, the impacts are significantly increased in areas with high population density in vulnerable locations. The controls over impacts of landslides and avalanches are shown in Table 12.

Table 12 *Factors controlling the impact of landslides and avalanches*

Landslides	Avalanches
Land use is a dominant factor, especially where previously forested areas have been lost	Downward velocity depends on slope angle, density of the snow mass, the shearing resistance at the base of the side and the total length of the downward path
Water plays a major role in mudflow and debris flow formation and transport; heavy soils can lubricate basal planes, decreasing shear resistance	The number of fatalities will be temporally dependent, especially in tourist areas such as ski resorts

Managing the landslide and avalanche hazard

Modify the event

Measures taken to *prevent* landslides include:

- building restraining structures such as gabions and stone walls
- excavation and filling of slopes to level them
- improving groundwater drainage and diverting surface water away from gully areas
- erosion control such as rock armour, revetments and use of 'gunite' (a sprayed mixture of sand, cement and water)

Methods for the control of avalanches include:

- artificially triggering avalanches using explosive charges or shelling before the snow accumulates to dangerous levels (especially prior to the start of the tourist season)
- planting trees along known avalanche paths (deforestation has increased the incidence of avalanches in some places, e.g. Austria, in the last 200 years)
- snow fences to accumulate snow in areas not prone to landslide hazards

Modify vulnerability

Landslide vulnerability can be reduced by:

- diversion of roads and infrastructure from known active areas
- planning control to prevent development in landslide-prone areas
- evacuation warnings when a landslide is imminent (e.g. when heavy rain is forecast)
- banning logging on hillsides

Avalanche vulnerability reduction involves:

- avalanche zoning (e.g. in France, there is no construction in 'red' zones (high hazard), while 'blue' zones (moderate hazard) mean special building-construction codes, such as reinforced concrete and no windows on up-slope sides)
- use of warnings to restrict access to vulnerable slopes (both Europe and the USA have systems in place)
- educating skiers in risk assessment and likely occurrence of events

Modify the loss

Landslide insurance is possible in some parts of the world, although it is difficult to obtain in high-risk locations. Personal insurance is possible in the event of an avalanche incident.

***Figure 67** Landslide hazard and risk evaluation (World Bank, 2006)*

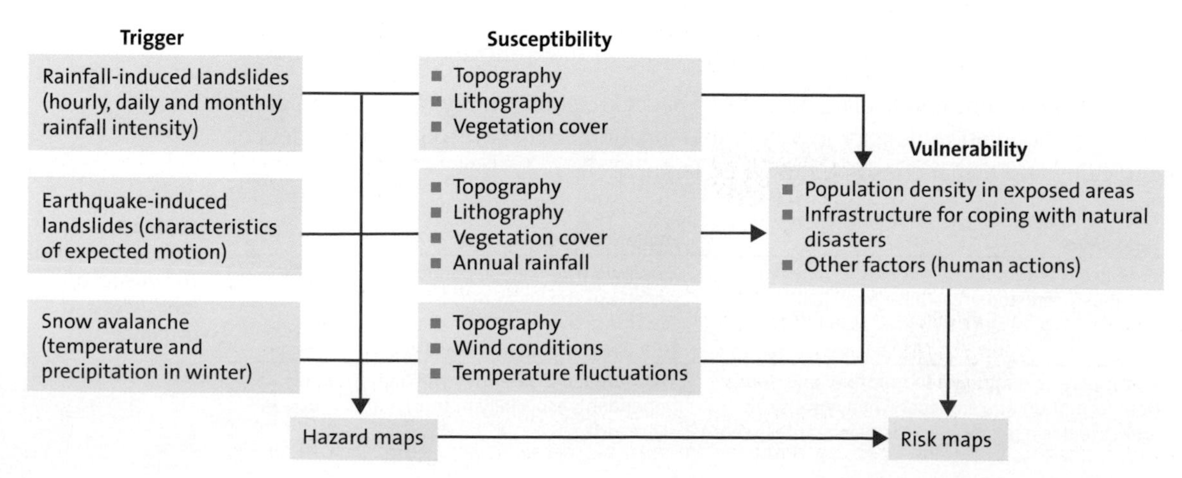

Computer models are becoming increasingly sophisticated, so that larger-scale risk maps can be produced. Figure 68 shows an example of a landslide risk map developed for Jamaica. Clearly, in areas of high risk, insurance is difficult to obtain, or is expensive for existing properties and businesses. Further development is therefore strictly controlled.

Figure 68* *Landslide risk assessment in Jamaica

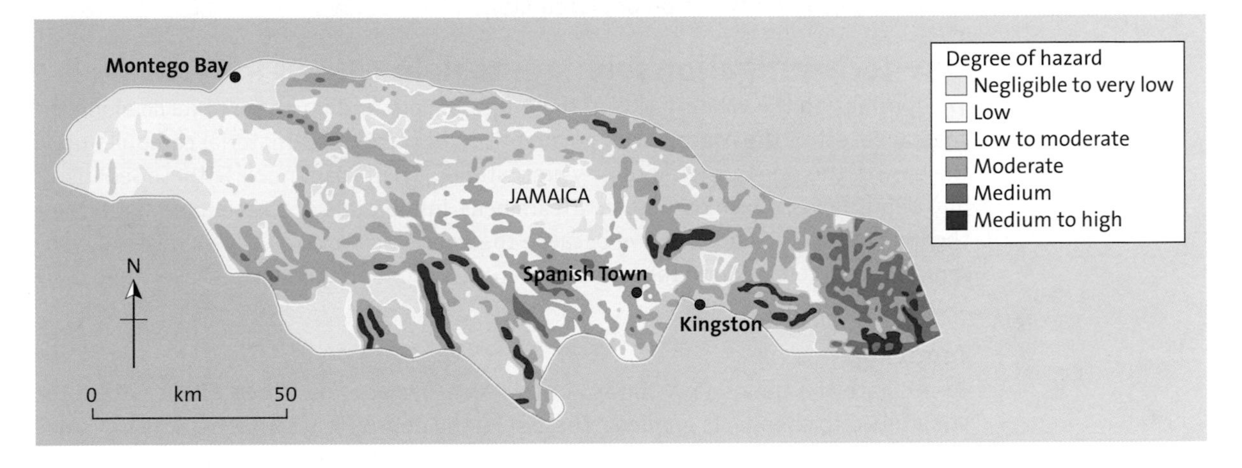

Summary

- Climate and weather conditions are crucial in determining the occurrence of both landslides and avalanches.
- Risk assessment, education and planning are decisive in terms of minimising impact. Post-disaster activity is a last resort.
- Current climate models for Europe and North America show increased winter temperatures leading to snowline rise. The same models predict increased heavy winter snowfall events at high altitudes in the collecting zones. Therefore, the risk of avalanche is increasing.
- Landslides may become more of a danger with increasing use of steep slopes for people to live on, combined with predicted higher-intensity rainfall patterns. Events such as debris flow, which occur infrequently, present particular problems since new migrants to an area often settle in hazardous areas, for example in shanty towns, and in *favelas* in Latin American cities, and are ignorant of the potential risk involved.

SOLUTIONS TO THE LANDSLIDE HAZARD

Technological solution to the landslide hazard in Hong Kong

Hong Kong is prone to landslides. In the 1960s and 1970s, management of the hazard was often reactive, that is, problems were sorted out after an event had occurred. But the late 1970s and 1980s saw a change from this approach to one that was more proactive — concentrating on landslide warning systems and prevention measures.

Systems were designed to give sufficient prior warning of an event so that the public and emergency services can be alerted. Warnings are based on groundwater levels (using an automatic piezometer) and measured rainfall. If the rainfall exceeds 175 mm in 24 hours, a landslide warning may be issued.

Prevention measures include cutting and filling slopes, and building retaining walls. This follows a detailed survey of areas at risk and prioritisation of work schedules. Slopes that are identified as at risk have a full stability analysis carried out, including shear testing and installation of piezometers.

Low-tech mitigation solutions to debris flows in Peru

The foothills on the western side of the Andes of Peru and northern Chile lie along the coastal desert of the Atacama. Large quantities of weathered material can build up at the base of the slopes. High-intensity precipitation events can mobilise these sediments into debris flows, which occupy dry stream beds. Inhabitants of the area may even think the dry stream beds are inactive because the debris flow events have a relatively long return period in relation to the human life span. Mitigation approaches follow two main low-tech solutions.

Terracing

To mitigate the debris flow hazard in this area, terraces have been built within the stream bed to change its gradient. These slow the debris flow through a series of small reductions in slope angle. Community organisation has been important in delivering this low-cost solution. Local inhabitants have constructed 24 such breaks with an investment of less than US$30 000. A similar approach has been adopted in California, where channels are built to divert flows as they approach settlements.

Land-use zoning

Land-use zoning to prevent the development of settlements in areas prone to debris events would reduce losses considerably. A lack of knowledge, and pressure on land, mean that such regulation is seldom introduced, despite the development of hazard maps at a local scale showing where the most stable slopes are and where there are high risks from debris flows and unstable hillsides.

11 *Using case studies*

Question

(a) Why have people become increasingly vulnerable to hazards that result from mass movements on slopes?

(b) Research an example of a widely known landslide event, such as Aberfan in south Wales in 1966, and then assess the statement that 'landslides are usually the result of both physical processes and human actions'.

Guidance

(a) People occupy and develop slopes, particularly in the crowded 'primate' cities associated with LEDCs. Landslides become more common, which increases vulnerability.

(b) Expect to write a 'thumbnail portrait' of your chosen event (location, date, number of deaths, damage caused). Then consider physical processes, which are likely to include rainfall problems, drainage issues, steepness of slope, structure and lithology, and the effect of human actions (basal steepening, overloading of upper part of slope, deforestation, and building in hazardous areas). Draw a summary diagram. Conclude by assessing whether you agree or disagree with the statement in the question.

Part 5

Biohazards

Biohazards can be classified as follows:

- Communicable diseases of humans (viral, bacterial, protozoal or parasitic), which can lead to local and regional **epidemics**, and have the capacity to turn into worldwide **pandemics** — for example, the Black Death in the fourteenth century, or the worldwide influenza outbreaks in 1917 and 1960.
- Epidemic crop diseases such as blights — for example, potato blight in Ireland, 1845–47, resulted in a famine that led to the deaths of 1.5 million people and caused a further 1.5 million to emigrate.
- Epidemic diseases of domesticated animals (bacterial or viral) — for example, Newcastle disease of poultry and foot-and-mouth disease of cattle and sheep.
- Epidemic diseases of wildlife and plant resources — for example, fungal diseases affecting cocoa pods.
- Plant and animal infestations — invasions of exotic pest species, such as locusts, which invaded the sub-Saharan republics in 2004–05, or the rabbit and cane toad plagues in Australia.
- Biomass fires, where spontaneous combustion by lightning in forest/grassland areas leads to wildfires, especially in periods of extreme dryness or very low humidity.

This section provides in-depth case studies of wildfires and locusts.

Wildfires

Wildfires, commonly known as bushfires (Australia) or brush fires (North America), are a natural process in many ecosystems. They are frequently used in conservation areas such as wildernesses (in New South Wales) and remote national parks (Yellowstone) as a necessary and beneficial tool of ecosystem management (Figure 69). This is a controversial technique as it can also be a major cause of wildfire hazard.

***Figure 69** The ecological cycle of burning*

Cause	% of fires	Cause	% of fires
Industry, e.g. forestry mismanagement	0.3	Lightning	5.6
Sawmills	0.3	Transportation	8.1
Smokers	1.5	Arson	8.4
Campers	1.7	Miscellaneous, known	9.7
Power lines	1.8	Burning off, legal	12.3
Rubbish tips	3.2	Burning off, illegal land clearing	15.3
Domestic, children	4.3	Miscellaneous, unknown	27.5

Source: Average of data held by the NSW Department of Bushfires Services

***Table 13** Principal causes of bushfires in Australia*

Wildfires can be classified as a biohazard because they can result from spontaneous combustion from lightning strikes. However, they are frequently induced either directly or indirectly by human actions (Table 13).

Fires are a major hazard if they occur in populated areas. They can cause widespread ecological destruction of rare animal species such as the lizards of Canford Heath in Dorset in the summer of 2006, or the orang-utans of Sarawak and Sabah in the Great El Niño forest fires of 1997–98. Large-scale fires can also be economically problematic.

The nature, intensity and rate of spread of a wildfire depends on the types of plants involved, the topography, the strength and direction of winds, as well as the relative humidity of the atmosphere. Some fires travel along the ground, others spread via the canopies of tree crowns. For this reason it is difficult to create a standard profile for wildfire (see Figure 1 on p. vii).

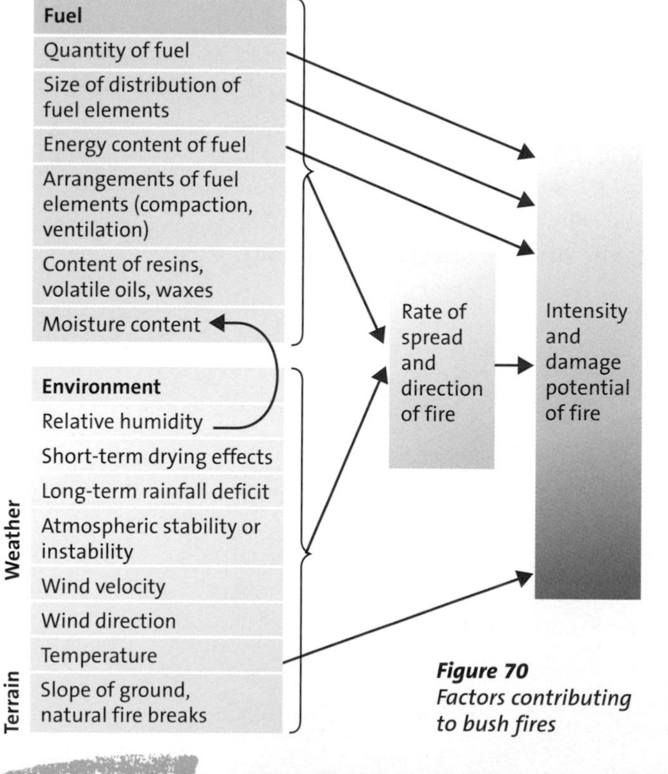

***Figure 70** Factors contributing to bush fires*

Wildfires are particularly associated with areas experiencing semi-arid climates where there is enough rainfall for vegetation to grow to provide fuel, yet with a dry season to promote ignition conditions (see Figure 70).

Wildfires are therefore concentrated in parts of Australia (NSW/Victoria), Canada (British Columbia), the USA (California/Mountain West and Florida), South Africa and southern Europe (in Mediterranean vegetation areas of Portugal, Greece, southern Spain and southern France). Traditionally, wildfires have not been associated with tropical rainforest areas because of the high humidity and all-year-round rainfall. However, burning for forest clearance and forest mismanagement by logging companies, combined with El Niño events leading to unusually dry conditions, have revised this.

Case study 17 INDONESIAN FOREST FIRES, 1997–98

The forest fires associated with the 1997–98 El Niño had the greatest impact in southeast Asia. They extended across the islands of Sumatra and Borneo (Sarawak) and generated a vast black cloud stretching over 300 km from west to east, engulfing several of the world's major cities such as Kuala Lumpur, Jakarta and Singapore.

Over 300 000 hectares of forest, often of high ecological value, were affected, scorched beyond recovery (Figure 71).

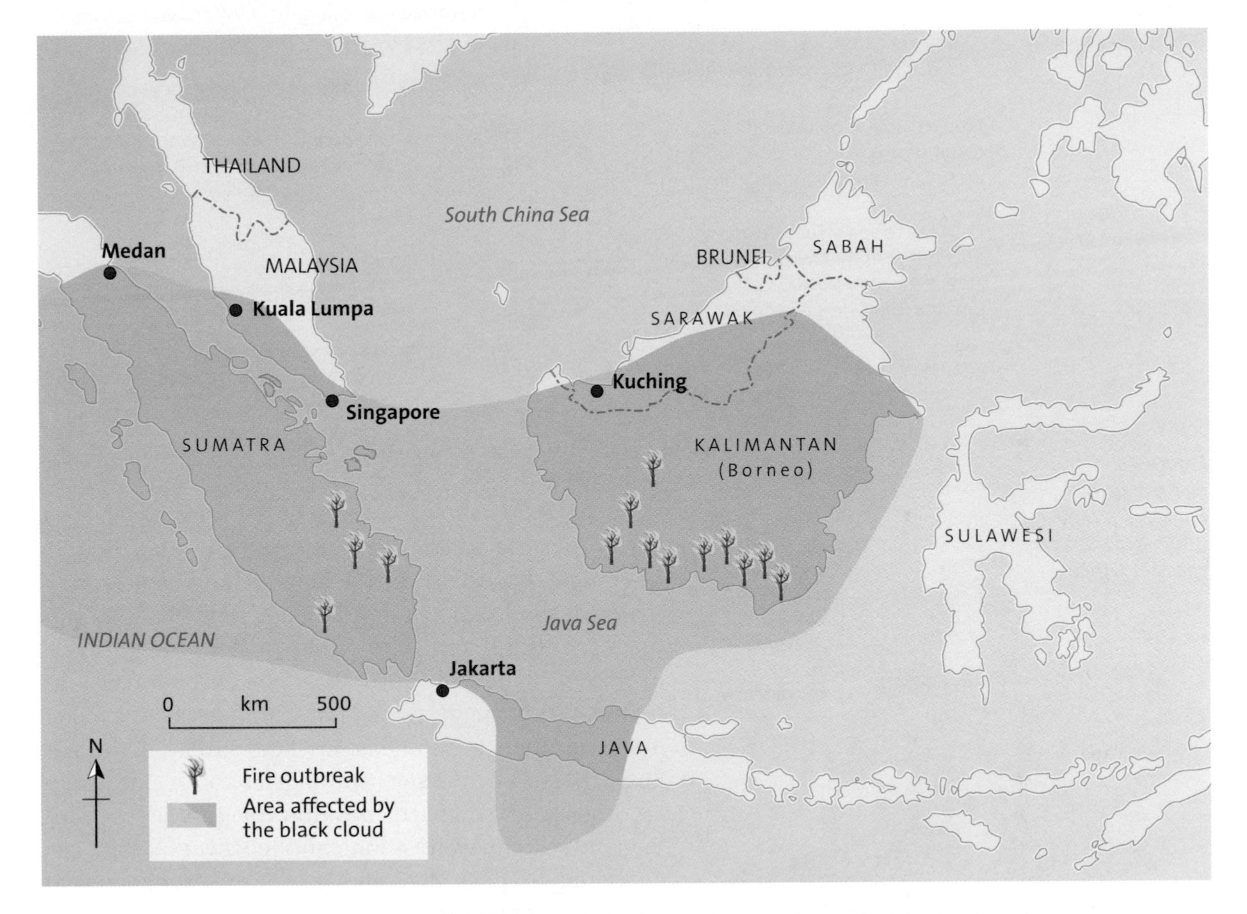

Figure 71
The extent of the black cloud produced by the Indonesian fires of 1997–98

The pollution from the fires produced a toxic haze. Around 60 000 people required hospital treatment for breathing difficulties.

The tourist industry suffered large losses for nearly 2 years. Tourist trips to places as far away as southern Thailand were cancelled. Some of the world's busiest shipping lanes and international airports experienced transport chaos.

The widespread distribution and large scale of these forest fires is clearly very different from more localised fires. There were at least 15 major outbreaks.

An increasing hazard

Although wildfires are associated with rural areas and usually cause only a limited impact, the movement of people into rural areas in southeast Australia, California, Florida and South Africa has spread the risk from wildfires. As many areas are hit by droughts from global-warming-induced climate change — for example, the summer droughts in Europe in 2003, 2006 and 2007 — there is concern that wildfires could become an increasing hazard for many areas, with major economic consequences. Owing to the nature of the hazard, deaths are anticipated to remain low, although there are major impacts on the local environment and on property.

Figure 72
The impact of wildfires

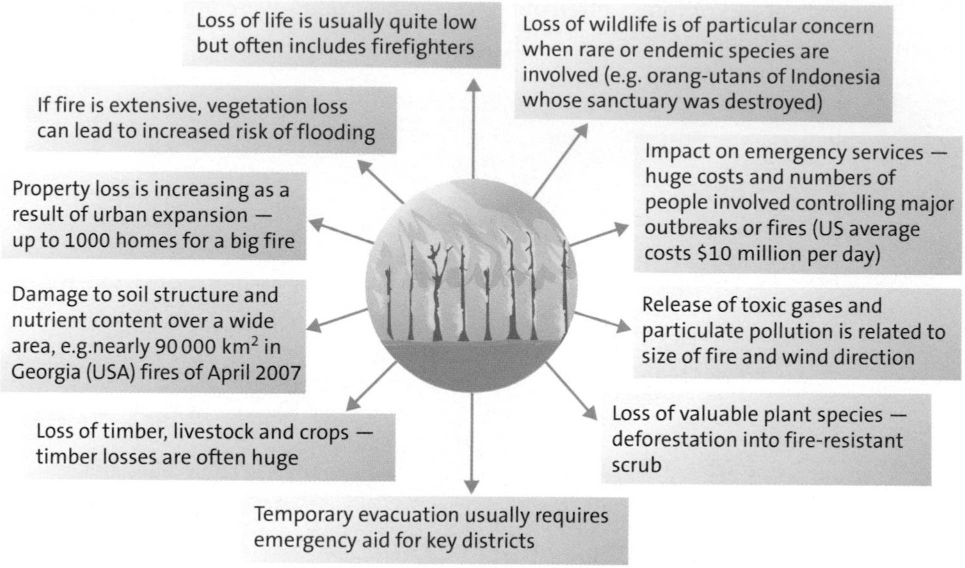

Loss of life is usually quite low but often includes firefighters

Loss of wildlife is of particular concern when rare or endemic species are involved (e.g. orang-utans of Indonesia whose sanctuary was destroyed)

If fire is extensive, vegetation loss can lead to increased risk of flooding

Impact on emergency services — huge costs and numbers of people involved controlling major outbreaks or fires (US average costs $10 million per day)

Property loss is increasing as a result of urban expansion — up to 1000 homes for a big fire

Damage to soil structure and nutrient content over a wide area, e.g.nearly 90 000 km^2 in Georgia (USA) fires of April 2007

Release of toxic gases and particulate pollution is related to size of fire and wind direction

Loss of timber, livestock and crops — timber losses are often huge

Loss of valuable plant species — deforestation into fire-resistant scrub

Temporary evacuation usually requires emergency aid for key districts

Figure 73
Fire risk across England, 3 May 2007, according to the Met Office

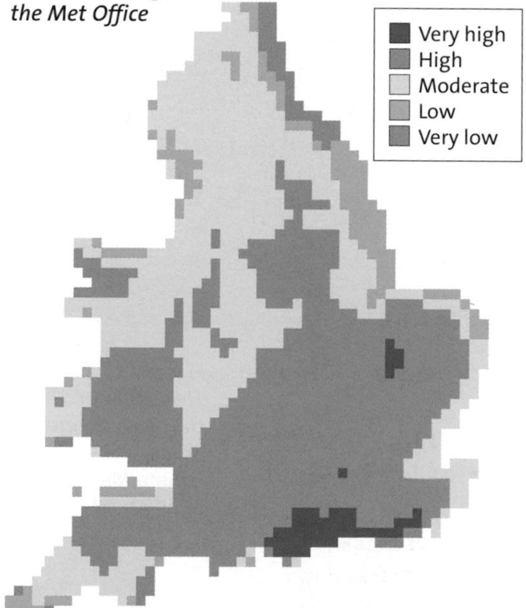

Forest fires are a hazard even in the UK, after periods of drought, especially for the ecology and environment — for example, the Canford Heath fires in 2005, and the recent North Yorkshire moors fires. Figure 73 shows the fire risk in May 2007 after an exceptionally dry period. With suggestions that climate change could make summers in southeast England similar to those in Spain, this may be a sign of things to come. Because of an unusually southern position of the jet stream, England had the wettest June and July ever in 2007 (see *Case study 6*, page 30), and it was the Mediterranean regions that experienced damaging forest fires, which killed 70 people in Greece.

'We need to shift our attitude to fire more in line with Californians, southern Europeans and Australians.' (Roger Ward, Natural England)

'It only takes a spark to cause a devastating blaze from which moorland habitats and wildlife take years to recover.' (Sean Prendergast, Peak District National Park)

Managing the wildfire hazard

Modify the event

In many countries, for example Australia, Greece and Spain, the main approach to fire management has been to extinguish all fires, especially in populated areas, or near to high-value timber reserves. Firefighters refer to the fire triangle — oxygen, fuel and a source of ignition being the apexes — when managing fires. With industrial fires, foam is used to exclude the oxygen supply. In a wildfire situation, however, oxygen is always abundant, so management to modify the event

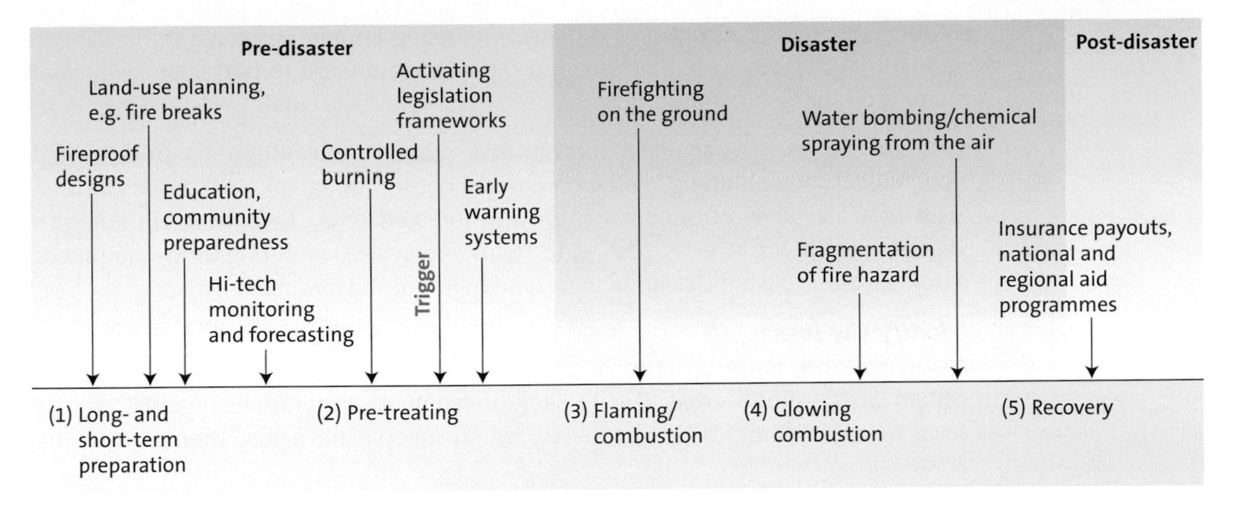

***Figure 74** Stages in development of a wildfire and its management*

concentrates on reducing or eliminating the fuel supplies from the potential path of the fire by controlled burning. This practice is highly controversial. It is a high-risk, polluting, labour-intensive measure that has a damaging effect on local ecosystems.

In national parks in the USA, such as Yellowstone, the removal of litter and lower vegetation is routinely used to minimise the possibility of unintentional renewed ignition. Prescribed burning is a widely used strategy in remote areas of the Mountain West.

Wildfire is one of the few hazards where management — using a combination of ground and air firefighting — can actually control and ultimately prevent the hazard becoming a disaster. Helicopters and planes act as water bombers and slurry (fire retardant) bombers. Where the benefits clearly outweigh the costs of control — for example, conserving large areas of housing in ever-sprawling suburbs — extreme management measures are taken to extinguish usually smaller-scale wildfires.

Modify vulnerability

A combination of the technological fix and community preparedness is vital in the management of wildfires. The technological fix is essential in remote areas to warn of fires. Aircraft and satellites are used to carry out infrared remote sensing to check surface temperature and signs of eco-stress from desiccation. In many US forests there are lightning detection systems, and infrared sensors; weather monitors and video cameras scan the forests of Florida for early warning precursors of wildfires.

Community preparedness can lead to early warning (using fire towers). Citizens can be trained as an auxiliary firefighting force, to organise evacuation and coordinate emergency firefighting. The 2007 Californian fires produced a model response.

Public education concerning home safety in high-risk areas is essential. Supplies of fuel should be reduced, wood stores stacked correctly, and adequate water hose and ladders should be available. All green waste should be composted and shrubs pruned regularly. Householders are also reminded to remove dead leaves from gutters. Warning levels of fire risk are also a vital component: high risk puts communities on alert.

School education concentrates on ensuring young people understand the dangers of arson and casual cigarette use, and the need to adhere to barbecue laws when there are high-risk conditions.

Land-use planning is again important. Risk management identifies areas of high vulnerability, and planning legislation ensures houses are built in low-density clusters with at least 30 m of set-back from any forested area. New developments are designed with fire breaks and wide roads for access of firefighting equipment. Fire-resistant housing design is increasingly important in areas of risk.

Modify the loss

Insurance is a common approach in MEDCs but it is expensive and difficult to obtain in fire-prone areas. Aid is likely to apply at a national, regional or local level in areas of hardship or poverty where people have lost their houses and possessions.

12 *Using case studies*

Question

(a) In your view, which of the three strategies (modify the event, modify vulnerability, modify the loss) is most important in responding to wildfires?

(b) Is it true that wildfires are a quasi-natural hazard?

Guidance

(a) Modifying vulnerability is the key strategy for minimising the impacts associated with wildfires. There is a growing risk for increasing numbers of vulnerable people, and increasing potential for periods of drought induced by climate change — although, unlike many tectonic hazards, the event can be modified by controlled burn. Modifying the loss applies to people who can afford the cost of insurance, which is inevitably high in areas of high fire risk.

(b) Spontaneous combustion by lightning strikes accounts for less than 10% of all wildfires. While certain natural conditions promote fire risk, vulnerability is clearly linked to the amount of settlement in a fire risk area. As Table 13 on page 62 shows, there are many human activities that increase the chance of fire. Most outbreaks involve a combination of triggers. On this evidence, quasi-natural is a fair classification of the wildfire hazard.

Locust plagues

The development of locust plagues is tied to locust breeding grounds in North Africa, the Sahel, the Arabian Peninsula and, periodically, parts of Asia, northwest India and Australia. Locusts are usually an unthreatening species. They breed rapidly and swarm only under favourable meteorological and environmental conditions:

- Rain promotes vegetation growth, which is necessary for locust development, and for the fat accumulation necessary for adult migration. It allows adults to mature and lay eggs, and speeds the development of eggs.
- Warmth promotes rapid development at all stages of the life cycle.
- Wind strength, direction and timing determine patterns of swarming and spread of locust plagues.

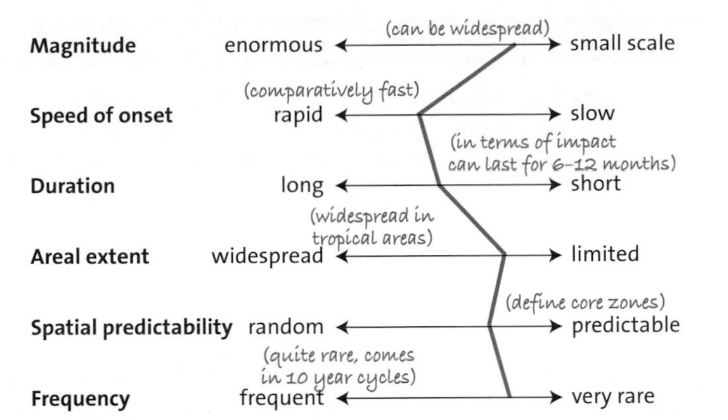

Figure 75
Typical profile of a locust plague

Topfoto

Figure 76
Locust plagues can spread from their breeding grounds as far as the Mediterranean and even across the Atlantic to the Caribbean; future climate change could lead to even wider spread

Managing the locust hazard

Modify the event

Control is the main option. Possibilities include:

- preventing the plague by controlling the upsurge stage — by spraying pesticide on breeding grounds
- eliminating the plague by destroying all, or nearly all, the locusts — this is an impossible mission, even by air-to-air spraying (where planes spray swarms in the air); it also involves environmental risks
- attempt to protect crops and allow the plague to follow its natural course

Modify vulnerability

The goal is to eliminate poverty, debt and food insecurity so that people can cope with a plague.

Modify the loss

Insurance is rarely an option but aid can be vital as shown by *Case study 18*.

LOCUSTS IN WEST AFRICA: EARLY WARNING, LATE RESPONSE

Warnings of locust plagues began to emerge in June 2003, after a very wet winter in northwest Africa. No vegetation escapes a hungry locust — a single swarm can eat the same amount of food in a day as the population of London will eat in a week. In spite of some efforts at pest control, strong northerly winds drove the locust swarms south to the Sahel, which also had a season of abundant rainfall (the first in years), causing the locusts to breed, especially in Niger and Mauritania (Figure 79). The Food and Agriculture Organization (FAO) issued numerous warnings of problems, but locusts bred while donors dithered. Subsistence farmers in these LDCs were completely helpless because of the size of the swarms and the lack of resources such as aeroplanes and fuel to spray pesticides on them.

The media gave little coverage to the crisis in west Africa — there were more exciting stories, such as the Iraq war. Agencies reported that appeals for food aid for west Africa raised nothing in many cases. It was not until October 2004 that war was declared on the locusts and the long-requested funds started to flow. Because FAO appeals fell on deaf ears, the infestation grew and spread, destroying millions of pounds worth of crops and pasture in the process. The costs of control escalated. However, by the end of 2004, a combination of control methods plus an unusually cold winter in the Atlas Mountains meant that the locust plague was under control. But that was by no means the end of the disaster, as millions of subsistence farmers struggled to survive a loss of crops and no pasture for dying livestock, thus contributing to widespread famine in 2005 (see *Case study 9*, p. 37).

Some of the world's most vulnerable people live in the Sahel. Around 10 million people face problems of recurrent drought, political instability (Niger), chronic poverty,

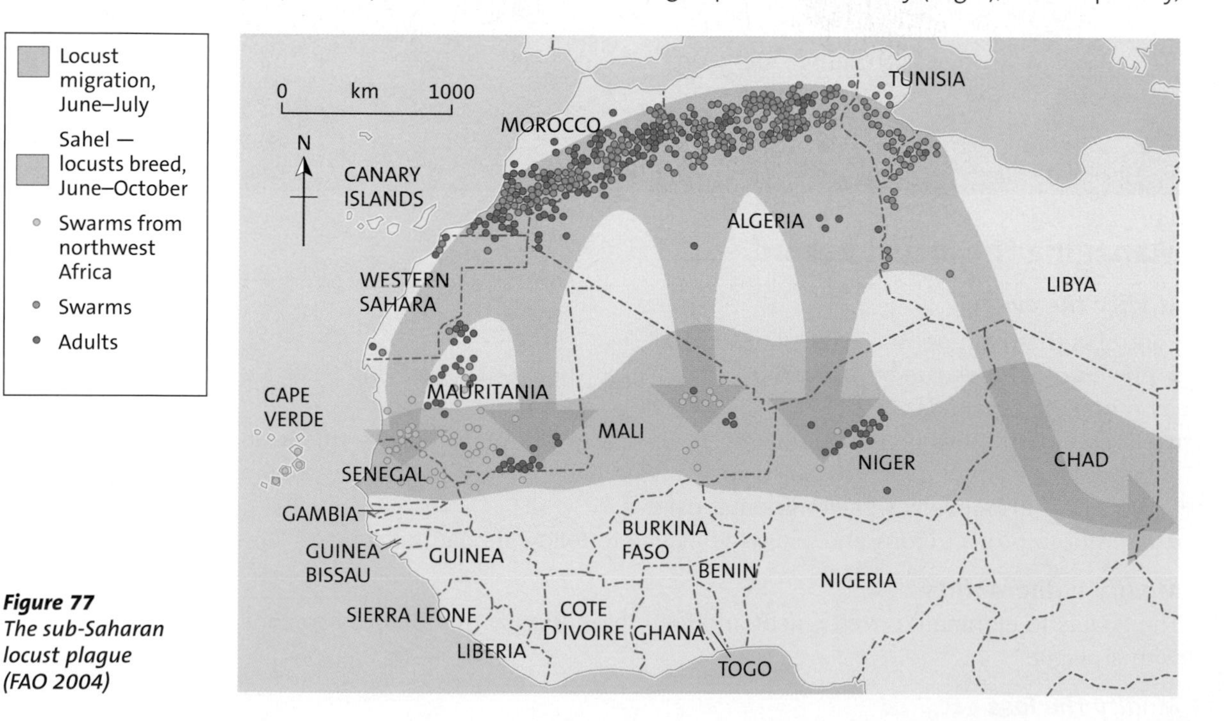

Figure 77
The sub-Saharan locust plague (FAO 2004)

national debt, poor health and now a lack of food security. The affected countries lack the resources to cope, and the regional cooperation required to fight transnational locust plagues proved difficult to generate. In the absence of recent experience of locust plagues, it is possible that international agencies underestimated the threat, especially the knock-on food crisis for nations 'living on the edge'. But there is no doubt that donations were too little and came too late.

Lessons that can be learnt from this disaster include the need for:

- properly financed and managed early warnings — always a problem for chronic and infrequent disasters
- better data to model the impact, so that aid agencies understand the scope of the problem
- stronger information campaigns to use the media to raise emergency funding
- a policy of spray now or pay later — early action would have avoided a 100-fold escalation of control costs
- greater cooperation at a regional scale to create greater awareness of common threats such as locusts and drought that lead to famine — while it is difficult to tackle the causes, it is often possible to raise awareness, issue early warnings and coordinate the response to the resultant food security issues

13 *Using case studies*

Question

What factors led to the locust hazard developing into a major disaster for the people of the Sahel?

Guidance

Look at *Case study 18* and make a note of the relevant factors — some are physical, most are human. Develop your answer along the following lines:

- the nature of the warning — failure to raise funding (late response)
- the nature of the hazard — perception, knock-on effects
- the nature of the location — LDCs, subsistence farmers, vulnerable people

Part 6

Multiple hazard zones

Identifying and defining hazard hotspots

In 2001, the World Bank's Disaster Management Facility (now the HMU, Hazard Management Unit) together with the Center for Hazards and Risk Research at Columbia University, USA, began a project to identify the world's key hotspots. These hotspots are **multiple hazard zones**.

The project assessed the risk of disaster mortality (i.e. death) and economic losses. Risk level indicators were estimated by combining exposure to the six major natural hazards (earthquakes, volcanoes, landslides, floods, drought and storms) with potential vulnerability for two elements of risk: mortality (population, population density/% poverty levels) and economic damage (GDP per unit area). Hazard exposure was estimated using data for the last 30 years where possible.

Historic records were used to build up a picture of past hazard-specific mortality (social cost) and assessments of damage (economic cost). Economic cost proved difficult to assess due to fluctuating currencies and lack of quantitative data, because so much economic loss is not insured.

***Figure 78** Identification of a hazard hotspot*

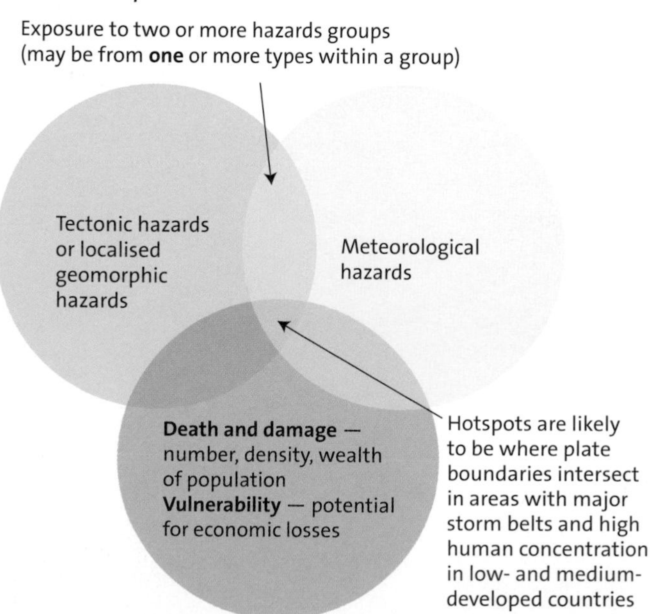

The World Bank's global risk analysis is noteworthy for its use of GIS techniques, which used grid squares 2.5 minutes of latitude and longitude as a base. Estimates of hazard probability, occurrence and extent were then related to the economic value of the land, the population, population density and vulnerability profiles. Since the aim was to identify disaster hotspots as opposed to hazard hotspots, only cells with a minimum population of 105 or densities above 5 per km^2 were entered on the database of around 4 million cells.

Data for analysing multiple hazard zones are often supplied by country, but data for small regions within a country, or within a city (see *Case study 20*, p. 75), are of much greater value for disaster managers and insurance industry risk assessors. Similarly, when answering questions on multiple hazard zones it is sensible to select data for a compact, more uniform (in terms of hazard risk and vulnerability) country, such as the Philippines or Sri Lanka, rather than, say, the USA.

The World Bank survey found that 3.8 million km^2 and 790 million people are highly exposed to two or more hazards; half a million km^2 and 105 million people are highly exposed to three or more hazards. Large numbers of people live in hazard-prone areas. Table 14 summarises some facts about multiple-hazard zones.

Figure 79 summarises the distribution of the major hazards and is annotated to identify some of the world's multiple hazard zones.

Table 14
Countries most exposed to three or more hazards

Country	% of total land area exposed	% of population exposed	Maximum number of hazards	Comment
Taiwan	73	73	4	Whole country vulnerable
Costa Rica	37	41	4	–
Vanuatu	29	21	3	–
Philippines	22	36	5	Severe hazards, e.g. volcanoes, earthquakes, typhoons
Guatemala	21	41	5	Nearly 60% of land exposed
Ecuador	14	24	5	–
Chile	13	54	4	Population concentrated in hazard-prone area
Japan	11	15	4	–

Source: Center for Hazards and Risk Research

Figure 79
The world's largest and most rapidly growing urban areas in relation to earthquake zones, volcanoes, tsunami-affected coasts and windstorm hazards

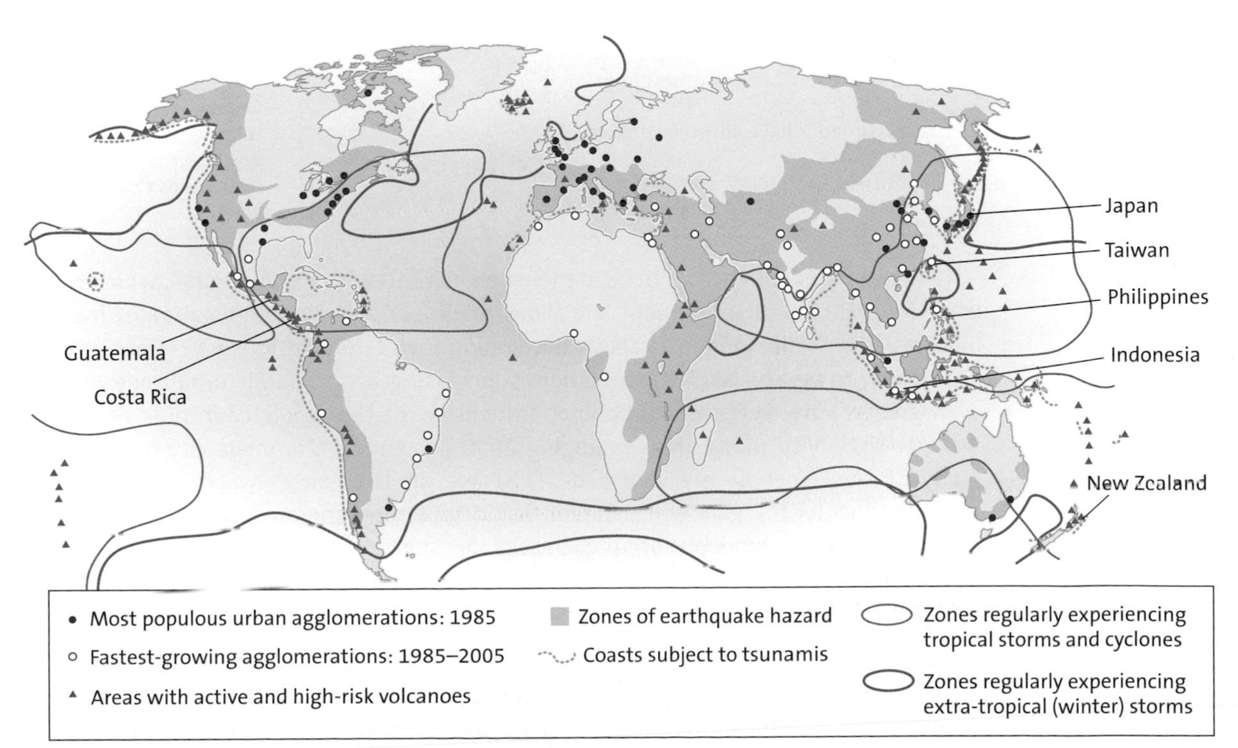

A CLASSIC MULTIPLE HAZARD ZONE

Figure 80 summarises the issues facing the Philippines, an island arc in southeast Asia. The country consists of over 7000 small islands, concentrated at latitudes between 5°N and 20°N of the equator. It lies within a belt of tropical storms (typhoons) and astride an active plate boundary. The dense oceanic Philippines plate is being subducted beneath the Eurasian plate.

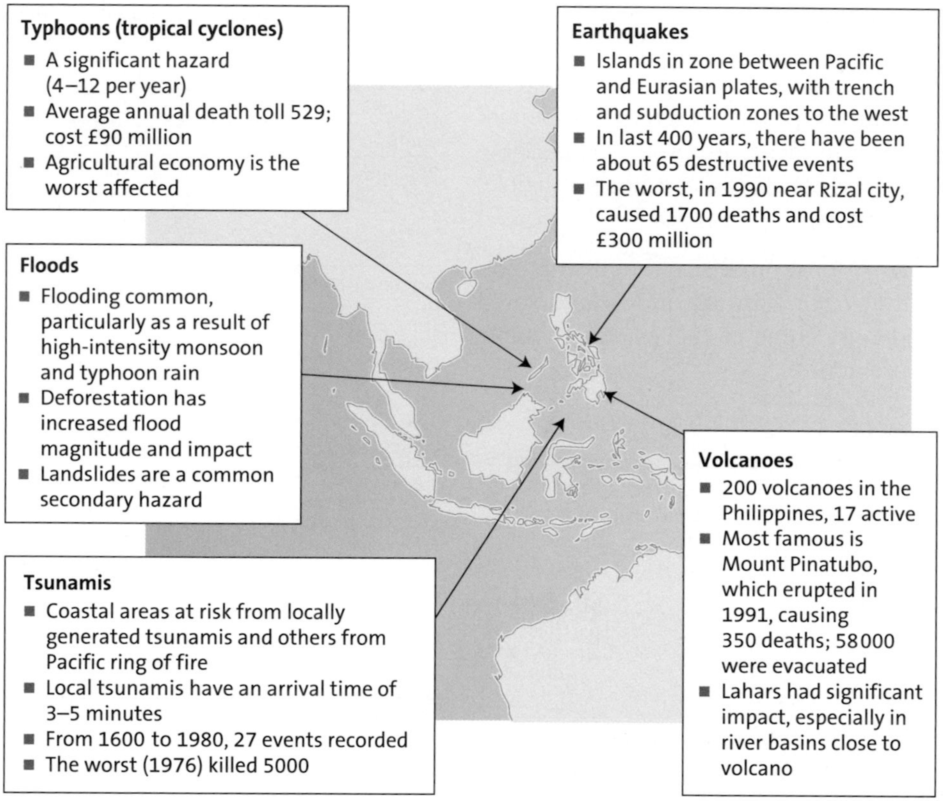

Figure 80 *The Philippines: a multiple hazard zone*

Typhoons (tropical cyclones)

- A significant hazard (4–12 per year)
- Average annual death toll 529; cost £90 million
- Agricultural economy is the worst affected

Earthquakes

- Islands in zone between Pacific and Eurasian plates, with trench and subduction zones to the west
- In last 400 years, there have been about 65 destructive events
- The worst, in 1990 near Rizal city, caused 1700 deaths and cost £300 million

Floods

- Flooding common, particularly as a result of high-intensity monsoon and typhoon rain
- Deforestation has increased flood magnitude and impact
- Landslides are a common secondary hazard

Volcanoes

- 200 volcanoes in the Philippines, 17 active
- Most famous is Mount Pinatubo, which erupted in 1991, causing 350 deaths; 58 000 were evacuated
- Lahars had significant impact, especially in river basins close to volcano

Tsunamis

- Coastal areas at risk from locally generated tsunamis and others from Pacific ring of fire
- Local tsunamis have an arrival time of 3–5 minutes
- From 1600 to 1980, 27 events recorded
- The worst (1976) killed 5000

The country experiences a tropical monsoon climate with heavy rainfall, which can lead to flooding and subsequent landslides made possible by deforestation of many hillsides. The Philippines is a rapidly developing lower–middle income country with aspirations to be an NIC. The population is increasing apace, and is urbanising at an even greater rate. The average population density for the whole country is high, at 240 per km^2, with many areas averaging 2000 per km^2 in the mega-city of Manila. Many of these people are very poor. They live on the coast, which makes them vulnerable to locally generated tsunamis and typhoon-generated storm surges. On average, about ten typhoons occur each season, especially in Luzon.

Creating a multiple-hazard diary

Use the internet to gather details of multiple hazard hotspots and create your own multiple hazard diary. Any of the top eight countries shown in Table 14 should provide similar results. The following list is an example of a multiple-hazard diary for hazard events in Japan in 2004 (Figure 81):

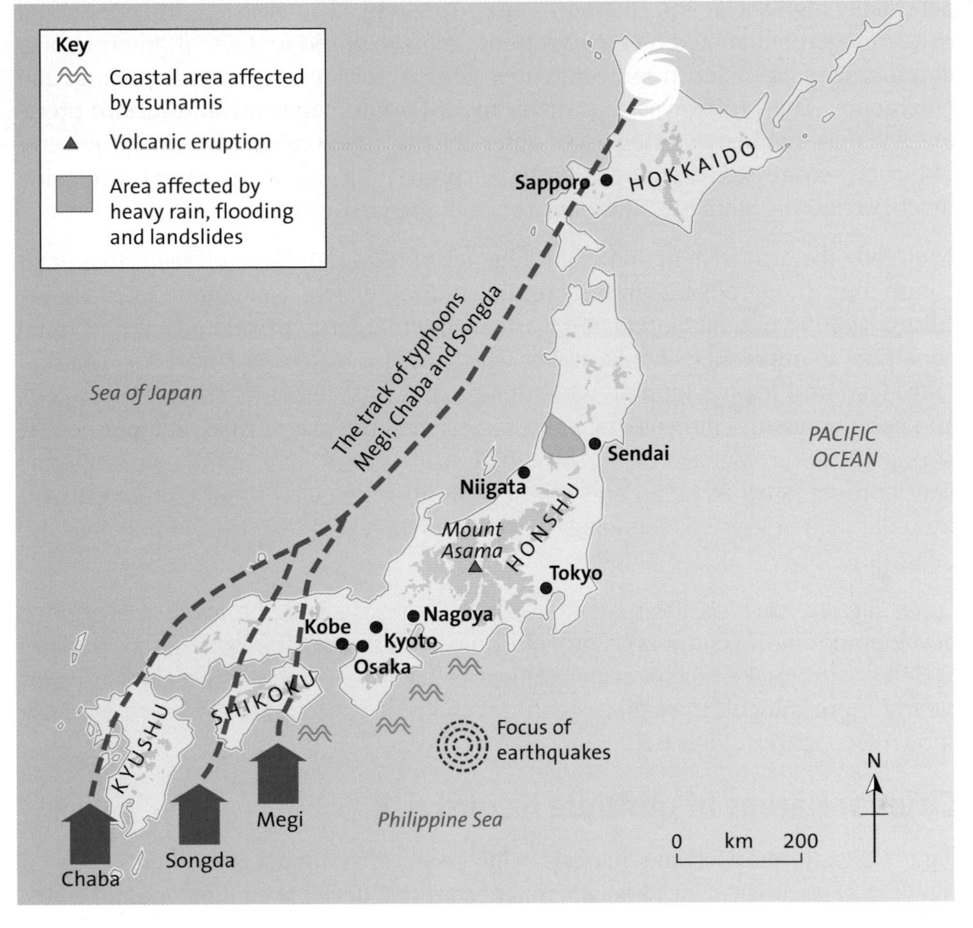

Figure 81 *Hazard events in Japan, 2004*

- 19 July: **flooding** — 18 dead; 490 mm rain (about 20% of total annual rainfall) fell in 4 days in Niigata state
- 19–20 August: **Typhoon Megi** — 90 mph winds; 205 mm rain; localised landslides and flooding
- 1 September: **volcanic eruption** — minor eruption of Mount Asama; no evacuation necessary; 183 minor earthquakes associated with the event
- 1–3 September: **Typhoon Chaba** — nine dead; 204 injured; 19 000 homes flooded; winds up to 130 mph damaged property and crops
- 6 and 8 September: **earthquakes** — 6.9- and 7.3-scale earthquakes hit southwest offshore Japan; depth and distance from coast reduced impacts; 1 m tsunami
- 7–9 September: **Typhoon Songda** — 225 injured; 83 000 households evacuated; 1.25 million people without electricity; wind speeds of 134 mph in southwest Japan

Managing a multiple hazard zone

The identification of multiple hazard zones has major implications for development and investment planning, and for disaster preparedness and loss prevention. Hazard events are expected to occur more frequently with global warming — and with potentially high social and economic costs. Theoretically, risk reduction efforts and risk management should be greatest in the most exposed areas, with international development agencies prioritising their efforts for contingency financing and emergency. The problem is that many hazard-prone countries and hazard-prone areas within them have a long list of other equally urgent priorities, such as poverty reduction strategies, fighting HIV/AIDS, promoting education, and achieving macro-economic stability. These are zones of greatest overall vulnerability.

Therefore, the attractive notion that frequency of hazard occurrence leads to greater community preparedness and decreased mortality does not apply. Even in the storms of 2004 that hit Florida, familiarity with emergency procedures almost 'bred contempt' in areas such as Lake Wales, which was hit by four hurricanes — Charley, Fran, Ivan and Jeanne. Coping with multiple hazards demands complex solutions and compounds the difficulty of a satisfactory programme of management before, during and after each event. In some high-risk regions with significant economic development, such as Japan and Los Angeles, it is also vital to protect investment from damage or loss, so technological solutions such as disaster-proof buildings are important.

Other factors, such as the magnitude of each hazard and the level of economic development of the community, are of prime importance. The Boxing Day tsunami in 2004 (earthquake Richter scale 9.1) was of such magnitude and scale that it was clearly more difficult to manage than the chain of varying hazards such as those shown for Japan in Figure 81.

Cities are areas of multiple hazard risk

Figure 82 summarises the reasons why large urban areas are often zones of multiple hazard risk. Cities are centres of economic development (economic cores)

Figure 82
Factors that make large cities vulnerable to disasters

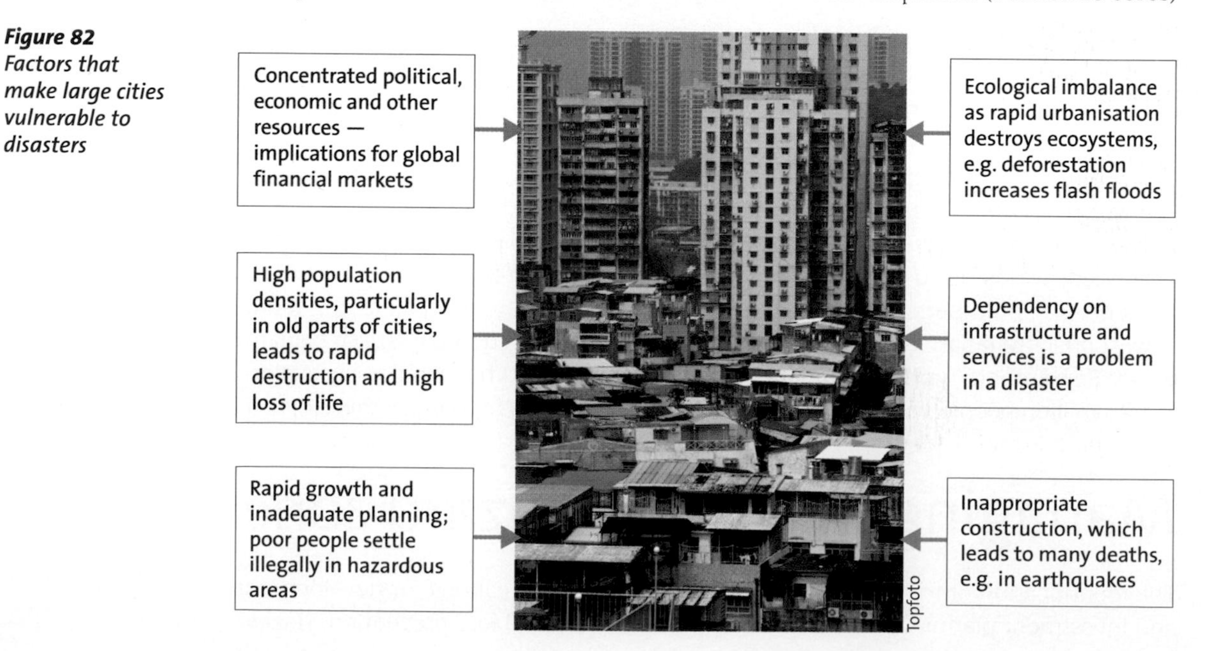

as they represent a natural focus for investment and development. They are also frequently centres of growing population as a result of the rapid urbanisation occurring in most developing countries. Many cities have huge areas of unplanned, poor-quality housing where growing numbers of the urban poor live, often located on marginal, potentially dangerous sites such as river banks and steep slopes. Analysis of the global distribution of these rapidly growing mega-cities shows that many of them are located in hazard-prone areas. With such high densities of people, up to 25 000 per km^2, hazard management in large urban areas is both expensive and complex, making disasters inevitable, both socially (high concentration of vulnerable people) and economically (huge investment in infrastructure, as in Tokyo).

As *Case studies 20* and *21* show, there is a 'collision' between the built environment and the prevalent natural hazards in the urban context.

CARACAS

Case study 20

Caracas, the capital of Venezuela, was unusually built some kilometres from the coast, on relatively steep land, as defence against sea-borne attacks. However, both Caracas port and airport are located on the Vargas coast and are reached by a highway that crosses landslide-prone valleys. As a primate city of nearly 6 million people, Caracas has experienced rapid urbanisation (nearly 4% growth per year). Many *barrios* (informal squatter settlements) have developed on the flanks of low-lying, yet rugged, mountains to the west and east of the city centre, where rainfall-induced debris flows are common. Between 15 December and 17 December 1999, 900 mm of rain fell in 72 hours, leading to landslides, mudflows and debris slides on the north face of El Avila. An estimated 25 000 people were killed. This event was the worst of many similar ones.

Caracas is also in an earthquake zone, located at the intersection of the South American and Caribbean plates. The last major earthquake in 1967 killed 300 people. The high population densities of 12 000 per km^2, and up to 25 000 per km^2 in the barrios, have led to widespread development of high-rise, densely packed apartment blocks, often located in identified zones of maximum potential earth shaking. The rapidity of the growth has meant that many houses, especially in the barrios, are constructed of unreinforced masonry, making them vulnerable to earthquake damage. Uncontrolled building and unenforced building laws have meant that many residents live in multi-hazard zones of flooding, land slipping and potential earthquake damage.

14 Using case studies

Question

What are the key factors of the built environment of Caracas that have helped it to develop into a multiple hazard zone?

Guidance

A mixture of relief, climate and tectonic instability, combined with a rapidly growing poor population, forced people to settle in hazard-prone areas.

LOS ANGELES

Los Angeles, in California, USA, was established on the coastal lowlands of the Los Angeles basin. It grew rapidly as a result of the oil and film industries to become a huge mega-city, extending 80 km from west to east and with a population of over 16 million. The physical environment has contributed to Los Angeles becoming a multi-hazardous urban environment, as shown in Table 15.

Table 15
Los Angeles: a multi-hazardous urban environment

Hazard	Causes	Impacts
Earthquakes	A network of active faults, including the San Andreas fault, underlies the Los Angeles region	The soft basin sediments lead to rapid shaking — five major earthquakes have been recorded in the last 100 years, including the Northridge earthquake
River flooding	Winter storms, especially during El Niño years, lead to floods in the Los Angeles and San Gabriel rivers, exacerbated by deforested hillsides	Rivers are now heavily channelised but flooding can still take place, usually between October and January
Coastal flooding	The area around Long Beach, which is subsiding, is sometimes flooded as a result of heavy storms	Likely to be an increasing threat as storm surges combine with rising sea levels associated with global warming
Drought	Always a potential summer problem in Mediterranean climates	Exacerbated by lack of water supplies for the increasing population in Los Angeles
Wildfires/ bush fires	A major hazard as Los Angeles expands into rural areas, especially during the dry Santa Ana wind periods (e.g. the Californian wildfires in Autumn 2007)	Likely to be an increasing hazard as people move out to the hills in the Los Angeles fringes
Landslides/ mudslides	Landslides take place in heavy winter storms where hillsides have been burnt by wildfire and eroded; also a risk along the coast near Malibu and Santa Monica	A growing risk as climate becomes more unpredictable
Smog	Climate conditions combine with car pollution to generate photochemical smog, which collects in the basin	A mega-city hazard, especially in late summer and autumn

Using case studies **15**

Question

(a) To what extent are the hazards in Los Angeles interrelated?
(b) To what extent are the hazards entirely natural?

Guidance

(a) Storms and landslides, earthquakes and landslides, drought and wildfires, fog, smog and ozone concentration

(b) Hazards are largely natural — for example, geologic, geomorphic and also hydro-meteorological hazards — *but* smog is quasi-natural, as are many wildfires (see Table 15).

Part 7

Responding to hazards and disasters

Natural hazards pose a risk to human life, livelihoods and possessions. The response to hazards can occur at a range of scales, from the individual within the local community to regional, national and international level, and, for mega-events or context hazards such as climate change, at a global scale.

Figure 83 shows how the choice of response depends on a complex and interlinked range of physical and human factors. As people and organisations have limited resources and time to make decisions, the relative importance of the physical risk from natural hazards, compared with other priorities such as providing jobs, education, health services and defence, will be a major factor in influencing how much resource is devoted to reducing hazard impacts.

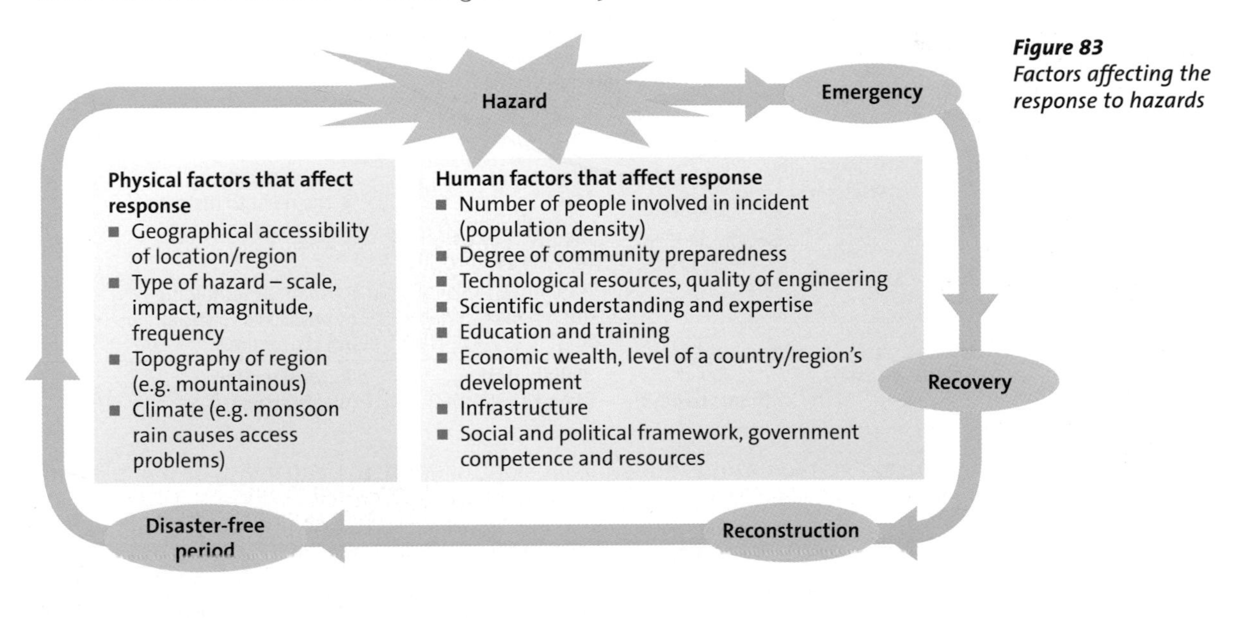

Figure 83
Factors affecting the response to hazards

When taking an overview of hazard events and their ability to develop into disasters, it is important to develop a number of frameworks, so that descriptive accounts of suffering and damage are avoided. Figure 84 summarises a useful framework for response analysis, which has been used in parts 2–5.

Modify the loss
- Aid vital for poor people
- Insurance more useful for people in richer communities and countries

Modify vulnerability
- Prediction and warning
- Community preparedness
- Education to change behaviour and prevent hazards realising into disasters

Modify the event
- Further environmental control
- Hazard avoidance by land use zoning
- Hazard-resistant design (e.g. building design to resist earthquakes)
- Engineering defences useful for coastal and river floods
- Retro fitting of homes is possible for protection

Modify the cause
- Environmental control
- Hazard prevention
- Only really possible for small-scale hazards, landslides/avalanches and floods

Increasingly technological

Figure 84
Response analysis framework

The choice of strategy will vary during the different phases of a hazard, as shown in the model of a disaster-response curve (Park response model) (Figure 85). This is an attempt to model the impact of a disaster from pre- to post-disaster. It also considers the role of emergency relief and rehabilitation. With each hazard event, or in the case of a multi-regional event such as the tsunami in 2004, each area or country may have a different response curve.

Figure 85
A disaster-response curve

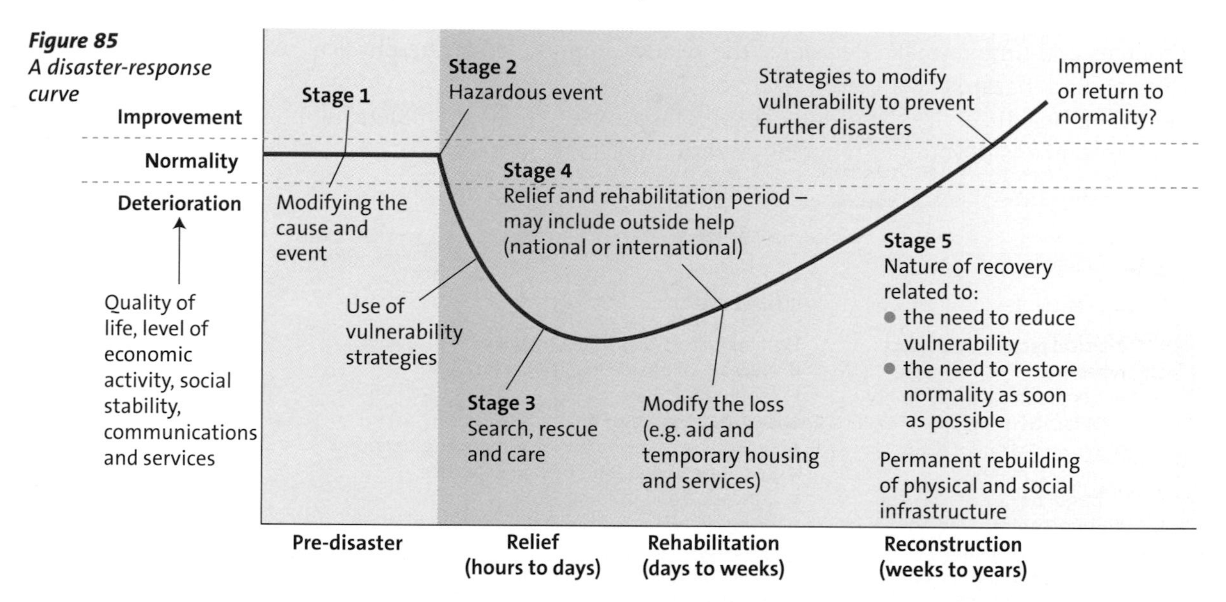

Table 16
The role of players in hazard management

The responses to hazards are critically controlled by the capability of the 'players' or groups involved in their management (Table 16).

Player	Role
International agencies and governments	Design and agree policies to manage disasters and to set up legislation, e.g. for emission control (Kyoto)
National government	National policies, civil protection and defence, public information; the Federal Emergency Management Agency (FEMA) was highlighted with Katrina; in the UK, the Environment Agency and DEFRA have responsibility for flood warning and management
Local government	Operational policies at the local level, e.g. rescue, welfare, medical services, transport, supply of emergency aid

Player	Role
Scientists, academics, educators, computer programmers	Researching, understanding causes, producing hazard (risk) maps, dissemination of information to the public — raising awareness, developing GIS
Insurers	Risk assessment prior to hazard; finance and assistance after hazard
Planners	Reducing risk by land-use zonation and planning, e.g. restricting development on floodplains
Relief agencies	Post-disaster aid, e.g. the International Red Cross featured during the 2004 tsunami
Engineers and architects	Design of buildings and infrastructure, ranging from Japanese technologists to US Corps of Engineers (rebuilding of New Orleans flood defences)
Emergency practitioners and services	Police, medics, firefighters, traffic control, possibly the army; coordination is an important role
Media	Highlight the hazard event (magnitude, scale, location, impact, duration); have a warning role that is important for more remote communities
Communities	Important role in managing situations, and in terms of community preparedness and education

THE ROLE OF PLAYERS

Case study 22

Kashmir earthquake (2005)

There was limited success in response to the earthquake in Kashmir in 2005 (see *Case study 1* on p. 8) because:

- the relief agencies (headed by the UN) failed to raise sufficient aid (US$550 million needed) to cope with the disaster — the USA was criticised for responding with too little and too late, which increased the number of secondary deaths
- in Pakistan, the army's initial response was slow
- the Pakistani government was reluctant to accept aid from Israel and India (because of a perceived political risk)
- the 80 000 death toll was so high because schools and hospitals were shoddily built
- much of the relief effort was organised by individuals who were originally from Kashmir and who wanted to help relatives, so effort was fragmented

The situation in Kashmir was particularly difficult to manage because:

- the area is politically unstable and is a war zone
- a disputed border meant that access from India was not allowed initially
- the area is isolated and inhospitable
- there is no electricity in some areas, and inadequate food and water; there is also a danger of disease spreading
- the incident coincided with the start of winter and the onset of very cold weather (made worse by the high altitudes)
- the numerous aftershocks hampered rescue efforts
- the magnitude of this disaster was so vast that the governments alone (especially in Pakistan and India) could not be expected to provide adequate relief

Boscastle flood (2004)

In contrast to the Kashmir earthquake, the response to the Boscastle flood (see p. 30) was successful. The timeline of events on 16 August 2004 was as follows:

- 2.00 p.m. — two Environment Agency (EA) operatives sent to check flooding of drains
- 3.15 p.m. — EA opens flood incident room in Bodmin

- 3.44 p.m. — local coastguard warns Falmouth coastguard of an incident developing at Boscastle
- 3.53 p.m. — five local firecrews sent to Boscastle
- 4.00 p.m. — visitor centre manager ushers families into the attic to escape the floods
- 4.45 p.m. — first of seven helicopters from Royal Navy, RAF and coastguard arrives
- 5.12 p.m. — fire and coastguard services declare a major incident and inform the media
- 5.23 p.m. — rescue helicopters begin winching people up from buildings
- 5.55 p.m. — Truro and Plymouth hospitals put on standby in case of casualties

The fact that there were no fatalities from this incident is largely due to the 'textbook' coordination of rescue and monitoring efforts from a range of agencies. A contributory factor may also have been the geographical situation of Boscastle — close to an RAF station with service personnel trained in rescue.

Hurricane Katrina (2005)

Hurricane Katrina exposed the shortcomings of local, state and national government. The mayor of New Orleans was critical of the coordination of relief efforts. The speed of relief was inadequate, causing unnecessary death and suffering. Katrina also cast doubt on the USA's ability to cope well in an extreme situation. The chaos that was Katrina aroused a model response and federal involvement for the Californian wildfires in autumn 2007, but not for the 2006 tornadoes.

Applying the Park model to Katrina

Ironically, the New Orleans event had been foreseen both in the short and long term. In the long term:

> New Orleans is a disaster waiting to happen. The city lies below sea level, in a bowl bordered by levées that fend off Lake Pontchartrain to the north and the Mississippi River to the south and west...the city is sinking further, putting it at increased flood risk after even minor storms. The low-lying Mississippi Delta, which buffers the city from the gulf, is also rapidly disappearing... A direct hit is inevitable. Large hurricanes come close every year.
>
> (*Scientific American*, October 2001, 'Drowning New Orleans')

In the short term, the hurricane-warning centre predicted a big hit close to the city days before the event.

On 28 August 2005, Mayor Ray Nagin ordered a mandatory evacuation of the city.

Impact of the disaster

The storm hit on 29 August 2005, with wind speeds of up to 190 km per hour (category 4). However, the real impact of Katrina was not the wind but the rising flood-waters (both inland and coastal). Biloxi recorded the largest ever storm surge (10 m) in the USA. On 30 August, a major levée near New Orleans failed and 80% of the city was flooded. An estimated 600 000 people fled the immediate area of the city.

The official death toll in Louisiana, Mississippi and Alabama was just over 1000. There were also significant economic consequences. Immediately after the event, the price of oil jumped to a record US$70 per barrel. The estimated cost of the disaster is believed to be about US$125 billion — a record for a single event.

Emergency relief

The American Red Cross quickly set up 275 shelters in nine states and supplied 249 emergency vehicles, 4200 relief workers and 140 000 meals.

On 31 August, government help started to arrive (organised by FEMA), amid much criticism — there were too many people on the Gulf coast when Katrina struck and too many were poor or immobile. The government did not provide enough help to get them out. The breakdown of law and order in the city was particularly disturbing. Armed police were ordered to tackle the lawlessness. Military transport planes evacuated only the seriously sick and injured to Houston. On 2–3 September, the military took over the city. Refugees were removed to other cities.

Rehabilitation and recovery

Katrina has highlighted the racial divide that still exists in the USA. Poor black people were in the most trouble; the wealthier white population had a better chance. By March 2006, less than half the city's former 450 000 residents had returned.

However, there has been a range of rehabilitation and rebuild events:

- February 2006 — Mardi Gras took place. It was needed to kickstart the local economy. New Orleans was once more open for business.
- March 2006 — FEMA released revised flood-risk maps. These have an impact on new building codes and standards as well as defining permissible locations.
- April–May 2006 — the Bring New Orleans Back Commission plan was launched. This is to help oversee the city's redevelopment and plays a role in deciding which part of the city is to be revived first.
- June 2006 — funds from the western Gulf oil and gas lease would help pay for restoration of Louisiana's marshes and barrier islands, which act as natural buffers. This was the deadline for completion of levée system repairs by the US army corps.
- September 2007 was the deadline for returning all levées and floodwalls to their original design height, although in the future, defences will have to be raised further because the city is subsiding.

In the longer term, coastal restoration will create a buffer against storms and will protect rural areas, ring-fencing the city with stronger levées.

16 Using case studies

Question

(a) Explain how and why the players found the Kashmiri earthquake difficult to manage.

(b) Why was the players' response at Boscastle so successful?

(c) Explain how shortcomings in the players' decision-making contributed to the disaster of Hurricane Katrina before, during and after the event.

Guidance

Use information from the case study outlined above.

The value of community preparedness

Preparedness is essential to ensure an effective response to a hazard event. It involves the detailed planning and testing of effective and rapid short-term responses, such as temporary evacuation plans, the preparation and distribution of emergency food and shelter, and the provision of medical aid. This is most effective at the community level, so it is important to train key people in first aid, search and rescue, firefighting etc., who then go on to educate the general population. In most events, 80% of the people saved are rescued by other survivors. International emergency preparedness can also be successful, via specialist rapid response teams, but there is inevitably a time lag. Much of this help is run on military lines, sometimes ignoring local perceptions and needs and the potential value of local resources (skills and knowledge). As a result, there are often political issues to resolve.

Community preparedness is even more effective when there is access to technology and it is combined with education initiatives.

COMMUNITY PREPAREDNESS

Avalanches in Norway

Snow and slush avalanches are a natural hazard to local communities in parts of Norway. They cause human fatalities and significant damage to houses and infrastructure every winter.

Geiranger is an area on the west coast of Norway with a high exposure to snow avalanches. Relocating 1000 residents is not realistic, so energy was devoted to finding acceptable means by which they could live, with minimised risks. The assessment concluded that any building mitigation measures could not be justified because of the high cost set against the low frequency of events. Instead, an early warning system together with a preparedness plan based on community actions was adopted:

- Get technical help to make a hazard map for avalanche-prone areas.
- Organise a local avalanche group consisting of representatives from community, including the police and civil defence agency.
- Install meteorological equipment to help with avalanche prediction.
- Develop an action plan for different hazard levels, including procedures for warning, evacuation and training of the local avalanche group.

The system was put to test on 4 March 2000. The hazard level was high and 32 people were evacuated to a hotel in a safe area. An additional 180 people were trapped between two avalanches because of an impassable road, but were successfully evacuated. Because of the well-developed preparedness plan, all operations were carried out successfully without the loss of life.

The learning experience from this case is positive. Several other communities along the western coast of Norway are adopting a similar approach.

Floods in Mozambique

The response to the floods in 2000 in Mozambique — the worst for a century — was in some ways a success. Of great significance were the 45 000 lives saved by rescue efforts coordinated and delivered by regional, rather than international, rescuers.

In 2001, another wave of devastating floods hit a different part of Mozambique. Local teams, operating mainly by boat, rescued over 7000 people. In each year, for every person who died, over 60 were saved. While media images of helicopters rescuing poor Africans gave the impression that international aid agencies saved the day, the real story is very different. Despite being one of the world's poorest countries, Mozambique was well prepared. Although international help was crucial, it succeeded because agencies let Mozambicans take the lead.

The success was due to a number of central factors, including:

- having safe places for people and cattle during an evacuation
- clearly marked escape routes
- legal powers that force people to move

Success was also attributed to simulations of major flood incidents involving the police, Mozambique Red Cross, local flying clubs, fire brigade and Scouts.

There has been criticism of the community preparedness scheme:

- Mozambique has to choose between disaster preparedness and other initiatives, such as healthcare and sanitation.
- There is an acute lack of financial resources — flood preparedness has dropped down the list of priorities.
- Much of the rainfall-monitoring infrastructure was destroyed in the 2000 floods and has not been replaced.
- Many of the lessons learnt were not realised in the 2007 floods.

Rebuilding communities after an earthquake in India

In January 2001, immediately after the Bhuj earthquake in Gujarat, Indian community-based organisations began to help in the recovery effort. A policy was proposed that would not only rebuild the devastated Gujarat communities but also reform and strengthen their social and political structures. The central concept was that people need to rebuild their own communities. Key elements of the strategy included:

- using reconstruction as an opportunity to develop local capacities and skills
- forming village development committees, made up of women's groups, to manage rehabilitation
- engaging village committees in monitoring earthquake-resistant construction
- striving to locate financial and technical assistance within easy reach of affected communities, and not being dependent on it being mediated by others
- distributing information about earthquake safety
- encouraging the use of local skills and labour, and retraining local artisans in earthquake-resistant technology
- encouraging coordination among government officials, district authorities and NGOs and seeking to facilitate public–private partnerships for economic and infrastructure development

This example shows how community involvement had added value — not just by contributing to the rebuilding and rehabilitation of the communities, but also in strengthening relationships between local stakeholders and the government and in empowering and engaging women.

Role of education in hazard management

The impacts of past events have revealed the important context of education for disaster-risk reduction: children who know how to react during an earthquake, community leaders who have learned how to warn their neighbours in a timely manner, and societies familiar with preparing themselves for natural hazards all demonstrate how education can make an important difference in protecting people at the time of a crisis. Education, combined with community preparedness, should prevent another major tsunami disaster, even if the techno-fix fails. In Bangladesh, in November 2007, this proved successful in reducing the number of deaths from cyclone Sidr, where 3000 people died, compared with cyclones in the 1960s, which claimed 250 000 lives.

Education for dealing with risk and disaster preparedness requires long-term investment in both MEDCs and LEDCs. Cultural norms and values, as well as related risk perceptions, must shift — a process that cannot happen overnight. Education requires a constant and consistent approach, beginning at an early age and continuing through generations.

Priorities of disaster-risk education must be embedded into education and training programmes:

- proceeding beyond a consideration of emergency response
- incorporating risk education in national development programmes
- educating about the social dimensions of risk in order to combat vulnerability

Some countries have dedicated emergency management agencies (e.g. the Federal Emergency Management Agency (FEMA) in the USA); in other cases, existing bodies, such as the fire and police services, deal with emergencies. Although preparedness is less likely to be available in LEDCs, it does exist (as the examples in *Case study 24* show), and many strategies are applicable across the development divide. For many poor communities, preparedness is the most effective, low-cost way to prevent disasters.

The successful operation of emergency disaster plans depends on public understanding and local cooperation. Cuba has some of the most effective hurricane management plans in the world, largely because public cooperation is mandatory in communist countries.

Education programmes can operate as bottom-up strategies (Nepal) or alternatively be developed within a global framework (UN).

EDUCATION AND TRAINING AT LOCAL AND GLOBAL LEVELS

Practical training in Nepal

Nepal is becoming more vulnerable to earthquake risk because of the increasing population, uncontrolled urban development and construction practices that have deteriorated over the last century. Despite this growing risk, until 1993, there was no institution concerned with this issue. Then, the National Society for Earthquake Technology (NSET) was established to confront the problem.

NSET has been involved in national and international projects, including one in Kathmandu that led to the development and implementation of the Kathmandu Valley Earthquake Risk Management Action Plan in 1998. More than 100 engineering students participated in a building inventory and vulnerability analysis programme. In particular, they were involved in aspects of safer construction in earthquake-prone areas.

A particular benefit of this programme has been the spread of information to local communities.

UN Disaster Management Training Programme (DMTP)

The DMTP is a global programme, which supports capacity-building efforts through international organisations and individual disaster-prone countries. Workshops have promoted the establishment of national or regional centres and strengthened their capacities to study technological and environmental hazards, seismic protection, crisis prevention and preparedness. DMTP has conducted more than 70 workshops involving 6000 participants in Africa, Latin America and the Caribbean, Asia and the Pacific, the Middle East and the Commonwealth of Independent States.

The DMTP provides professional and structured learning and skill-building programmes that also serve to:

- support and create synergies among partner organisations
- raise the profile and visibility of disaster management in areas of particular risk
- promote awareness-raising and motivation, adaptability, increased ownership and responsibility
- encourage the commitment of people, local and international resources, technologies and funding

The technological fix

In theory, technology should reduce the impact of natural hazards at all stages of the disaster management cycle (Figure 86).

The techno-fix has come to the forefront of disaster management due to the development of a range of new technologies, including geographical information systems (GIS), global positioning systems (GPS) and remote sensing (RS). Improved computer processing and storage provide opportunities to develop project planning and real-time decision making in emergencies (e.g. damage assessment and evacuation routes).

The increasing reliability of portable radio-based transmission systems and satellite phones allows communication even when ground-based infrastructure is inoperable or damaged. GPS technology is now much cheaper and widely available, as well as being portable. Its use with notebooks or handheld computers allows accurate position fixing and direction finding in remote areas.

Remote sensing in disaster reduction

Monitoring is essential for the detection of the onset of disasters. It also helps in appraising a disaster situation and in effective recovery and reconstruction. It is often necessary to monitor vast areas; airborne observation systems, including satellites, are the most effective for this purpose.

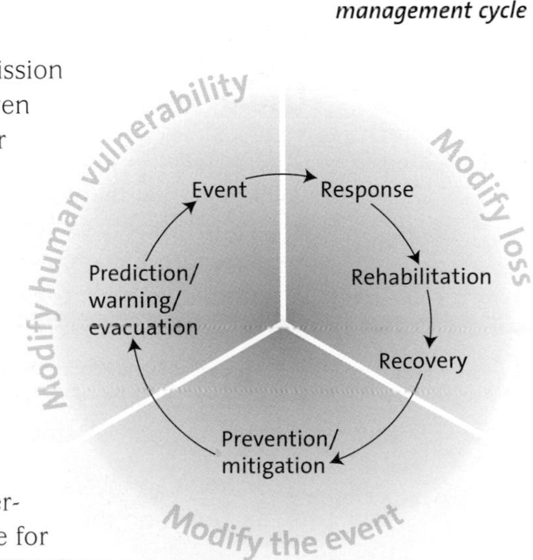

Figure 86
The disaster management cycle

Applications of remote sensing (RS) in disaster reduction and management include:

- **cyclone detection and warning**. Meteorological satellites play the main role in detecting and tracking cyclones as they form and move over the ocean.
- **weather radar**. A weather radar provides a quick estimate of rainfall density covering a large area. The continuous measurement of rainfall movement has been used to develop predictive models for short-term rainfall forecasts.
- **drought monitoring**. RS data cannot be used for early warning of drought, but they can be used to appraise the extent of drought impact. At national level, the main objective is to secure food supply, mostly through international assistance. Monitoring is achieved through interpretation of satellite images.
- **earthquake damage mitigation**. Satellites provide general information on land use that would support various civil engineering works for mitigation and response. Interferometry synthetic aperture radar applications have been used to view small (centimetre-level) crust movements that have occurred during earthquakes. This technology has so far had limited use in earthquake prediction.
- **flood damage reduction**. Meteorological satellites and earth observation satellites can be used to estimate hydrological variables (snow cover, water elevation, rain rate, soil moisture, solar radiation, surface albedo, land cover, flood monitoring and surface temperature). The real strength of RS in flood mitigation is the supply of parameters that are difficult to measure for flood modelling, particularly data on land use and vegetation. This information is also required in the development of flood-risk maps.

Use of geographical information systems in disaster reduction

GIS provides a key resource for disaster mitigation, combining physical data (slope, geology, soils and drainage) with human and economic data (census, housing types and location of vulnerable people). GIS creates 'layers' of electronic data.

GIS has developed large-volume data handling capabilities, which facilitate information synthesis from different data sources and then provide readily understandable maps for hazard managers. Figure 87 summarises the role of GIS at all stages of the disaster management cycle.

Use of technology in prediction

The purpose of prediction is to allow people to respond at individual, community, local, regional, national and international levels. Accurate prediction can buy time:

- to warn people to evacuate
- to prepare for a hazard event
- to manage impacts more effectively
- to help insurance companies assess risk
- to prioritise government spending
- to help decision makers carry out cost–benefit calculations, for example for the construction of defences

Prediction can increase understanding, because scientists can model and test their predictions and then compare them with reality.

Figure 88 summarises the importance of prediction.

GIS is used for disaster recovery planning, in the short term to locate and restore vital services (secure lifelines, e.g. gas, water, electricity, and for resettlement safe houses), in the longer term to direct rebuilding and relief funding and to prepare recovery plans, visualising trends and tracking stages in recovery.

GIS is used to identify assets at risk and location of support facilities. Layers of information including land use, key facilities (e.g. schools and hospitals) and infrastructure are combined with physical characteristics, by administrative district, to create a disaster management system.

GIS provides support for mathematical simulation and modelling of disaster processes.

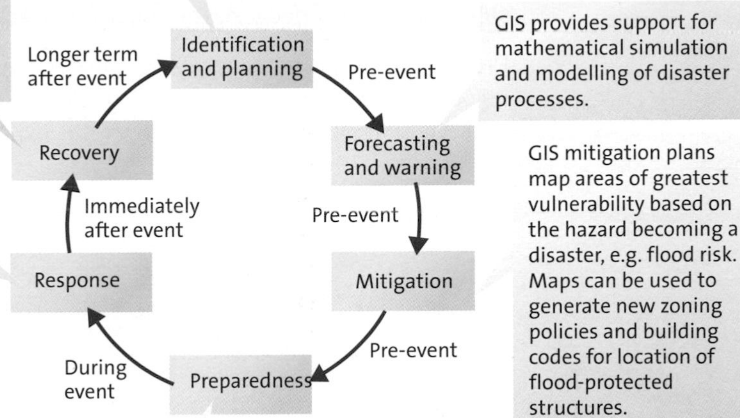

GIS is used for emergency management. Reconnaissance of the disaster and emergency mapping of the area affected are carried out, e.g. by helicopter equipped with GPS, flying in front of a fire at regular intervals, combined with other data, e.g. the weather forecast, to model spread and decide where to apply 'water bombing'. Damage assessment forecasts are based on tangible features such as buildings likely to collapse. Fragility functions for different land use areas can express potential damage as a percentage of cost, for a particular type of hazard, e.g. depth and duration of floodwater, or ground shaking intensity of earthquakes combined with fire risks, e.g. gas mains. This will direct scarce resources to likely areas of most emergency.

GIS mitigation plans map areas of greatest vulnerability based on the hazard becoming a disaster, e.g. flood risk. Maps can be used to generate new zoning policies and building codes for location of flood-protected structures.

GIS subdivides areas into management zones. Within each zone, it records the number, locations and types of support facility (e.g. sports stadiums/schools for emergency support, medical centre capacity, evacuation points, emergency response organisation, cemeteries/ morgues, availability of machinery etc., escape routes, back-up communication, location of vulnerable people). This information can be shared with the general public who need to know how warnings will be received, i.e. a disaster preparedness plan.

Figure 87 *Uses of GIS in the disaster management cycle*

When?

- **Recurrence intervals** — an indication of longer-term risk
- **Seasonality** — climatic and geomorphic hazards may have seasonal patterns, e.g. Atlantic hurricanes occur from June to November
- **Timing** — the hardest to predict, both in the long term (e.g. winter gales) and the short term (e.g. time of hurricane)

Where?

- **Regional scale** — easy to predict, e.g. plate boundaries, 'tornado alley', drought zones
- **Local scale** — more difficult, except for fixed-point hazards, e.g. floods, volcanoes, coastal erosion
- Moving hazards — extremely difficult, e.g. hurricane tracking

What?

- **Type of hazard** — many areas can be affected by more than one hazard; purpose of forecast is to predict what type of hazard might occur
- **Magnitude of hazard** — important in anticipating impacts and managing a response
- **Primary vs secondary impacts** — some hazards have 'multiple' natures; earthquakes may cause liquefaction, volcanoes may cause lahars

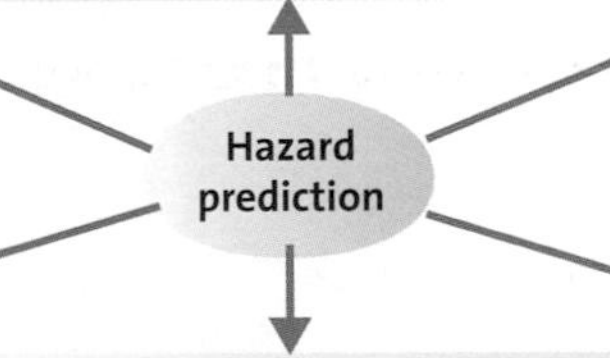

Why?

- **Reduce deaths** — by enabling evaluation
- **Reduce damage** — by enabling preparation
- **Enhance management** — by enabling cost–benefit calculations and risk assessment
- **Improve understanding** — by testing models against reality
- **Allow preparedness plans to be put in operation** — by individuals, local government, national agencies

Who?

- **Tell all?** — fair, but risks over-warning, scepticism and panic
- **Tell some?** — for example, emergency services, but may cause rumours and mistrust
- **Tell none?** — useful to test predictions, but difficult to justify

How?

- **Past records** — enable recurrence intervals to be estimated
- **Monitoring (physical)** — monitored and recorded using ground-based methods or, for climatic and volcanic hazards, remote sensing
- **Monitoring (human)** — factors influencing human vulnerability (e.g. incomes, exchange rates, unemployment); human impacts (e.g. deforestation)

Figure 88 *Hazard prediction*

There are sometimes problems with the reliability of prediction technology. False warnings cost huge amounts of money — for example, in Hawaii each false tsunami warning costs £30 million — and decrease confidence and credibility. People obeyed the first warning in the 2004 Florida hurricane season but were cavalier about subsequent ones because the hurricane tracks were unpredictable.

Prediction relies increasingly on sophisticated technology, so it is expensive to resource. As a result of the technology gap, an inequality exists between LDCs/LEDCs and NICs/MEDCs. Figure 89 shows the extensive range of methods available for predicting earthquakes.

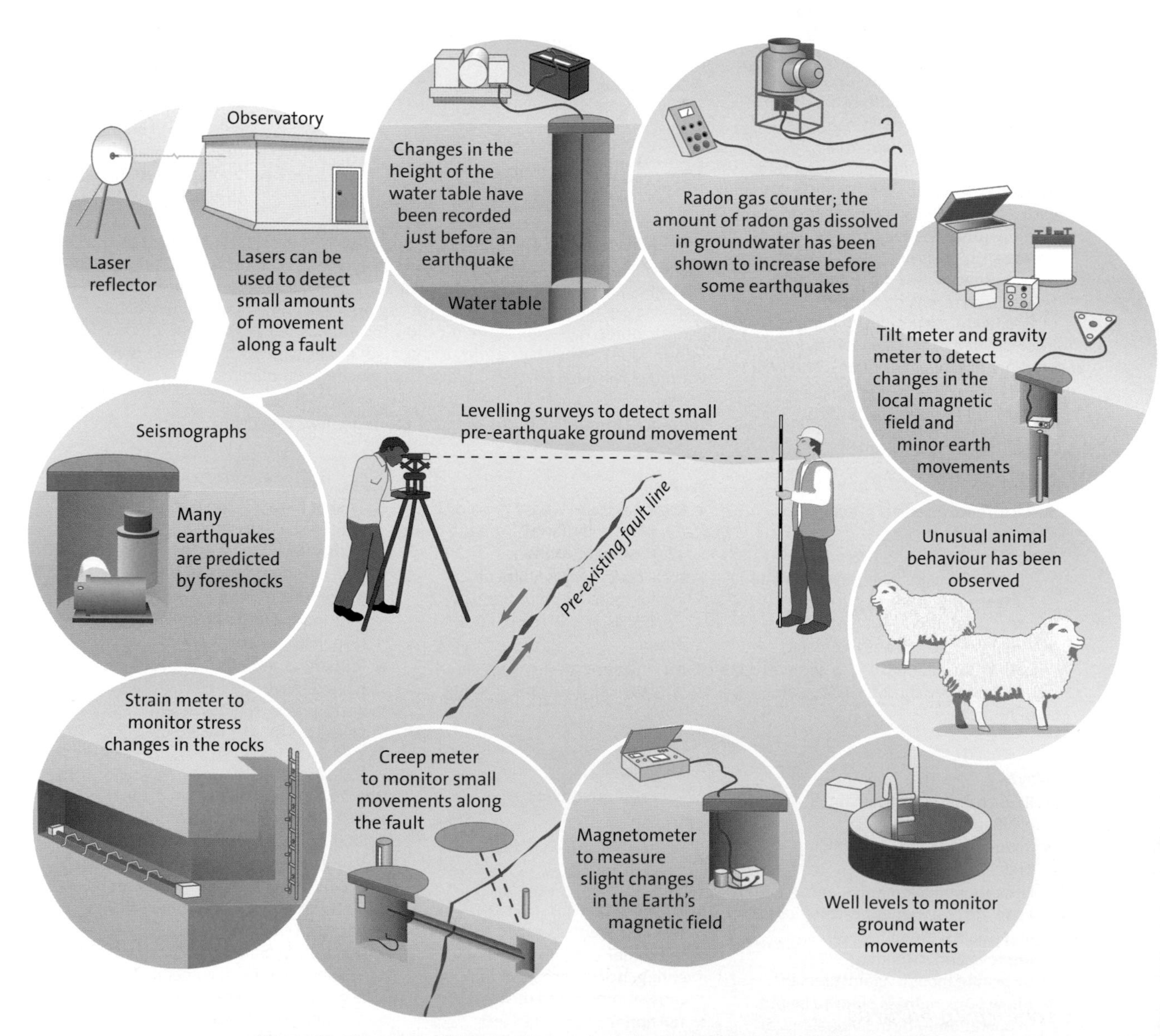

***Figure 89** The range of monitoring methods for predicting an earthquake along an active fault line — only some of these methods would be employed at any one site*

17 Using case studies

Suggest why, even in high-tech societies, high-tech equipment is not always successful in predicting hazard events.

Guidance

The shortfall of technology and its use:

- It is important to provide a framework for technologies at the national and local levels. For example, hazard maps or risk maps need to be distributed to, and used by, the community for which they are developed. This has to be done by local government through legislation and education programmes.
- It is not always feasible to have the technical capability locally (especially in NICs and LEDCs). However, in Mindoro in the Philippines, for example, the community enlisted the services of a consultant to prepare a hazard map for use in development planning.
- Standardisation of information is required. As information generated by one organisation is used by other organisations, it is necessary to have common standards.
- The basic information needs to be available to the general population. Researchers, industries and governments need to be encouraged to develop technology to counter disasters. The USA has taken the lead in making information accessible for the research community and the public. However, Japan has been slow to implement data accessibility, resulting in a waste of resources and time.
- Some technologies have different utility according to hazard type.

The nature of the hazard itself. The speed of onset (earthquake versus drought), lack of reliability of spatial behaviour (hurricanes), frequency and magnitude all play a part.

Modifying the hazard event

Causes of hazards are either **primary** (e.g. slope failure that triggers a landslide) or **root** causes induced by humans. Root causes are being managed increasingly as part of integrated risk-reduction strategies to modify vulnerability, but the management of primary causes is for the future. Managing the physical processes involves modifying the environment in some way so that physical exposure to the hazard is prevented. There are ethical issues in the more experimental approaches (humans controlling nature) and there is also the risk of adverse ecological and environmental consequences.

Tectonic hazards

Modifying and managing the causes of earthquakes is not feasible, although experiments with lubricating the San Andreas fault plane have been conducted to try to prevent the jarring that results from sudden slippage. Volcanic eruptions cannot be prevented. Attempts have been made to control lava flows by bombing fluid lava or by using artificial barriers to divert it (e.g. in Hawaii). On Heimaey in 1973, seawater from the harbour was used to chill the advancing lava from the Eldfell eruption. The causes of tsunamis are impossible to control.

Geomorphic hazards

Causal management of geomorphic hazards is feasible. For avalanches, snow packs can be stabilised. Artificial release occurs at predetermined times when ski runs and highways are closed. The snow pack is released safely in several small avalanches, rather than allowing a major threat to build up. Together with monitoring snow stability, small explosives are used to trigger controlled avalanches. Where highways are endangered, for example Rogers Pass in the Rockies, the armed forces may trigger avalanches with field guns. There are defence structures to control the development of wet-snow avalanches, but powder avalanches are difficult to manage.

Landslides and other slumping can also be prevented by slope defence methods such as excavating and filling, draining, re-vegetation, restraining structures and chemical stabilisation (e.g. in high-risk urban areas of Hong Kong).

Hydrometeorologic hazards

Suppression of severe storms remains a dream. A 10% reduction in wind speed could downgrade a tropical cyclone to tropical storm status and, therefore, perhaps reduce damage by 30%. In the 1960s, Project Stormfury aimed to introduce freezing nuclei into the ring of clouds around storm centres to stimulate the release of latent heat to lower the maximum temperature in the hot core (the programme was discontinued 1980s). Cloud seeding has also been used to suppress hail in severe storms and is being used to manage potential rainstorms at the Beijing Olympics. Current research is concentrated on attempting to cool down the oceans that generate hurricanes by covering the surface with biodegradable oil, and on how to cause hurricanes to change course to veer them away from densely populated areas.

In theory, droughts can be prevented by artificial stimulation of rainfall by silver iodide seeding, but the clouds must have natural precipitation potential.

Causal management of floods is feasible, largely by attenuating the lag time — using dams and reservoirs, and afforestation of the upper catchment.

18 Using case studies

Question

Use the continuum framework below to grade the main hazard types in terms of feasibility of the techno-fix being used for causal management.

Impossible ←——————————————————→ Feasible

Aid and insurance

Aid

Humanitarian concern for disaster victims results in the flow of emergency aid from governments, NGOs and private donors. Disaster aid is used for relief, rehabilitation and reconstruction (Figure 90).

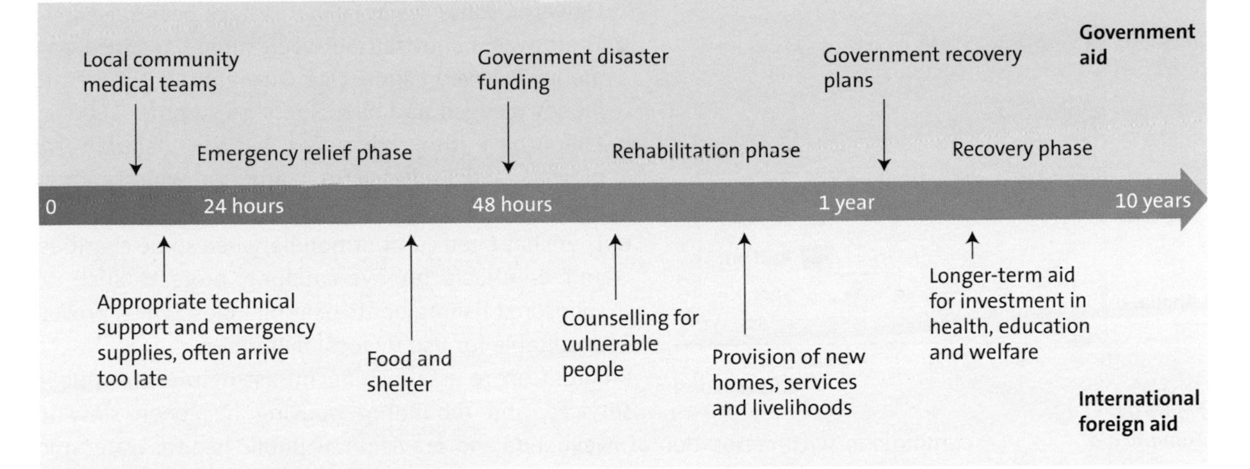

Figure 90 *Emergency aid — how it is used*

AID FOR THE 2004 TSUNAMI

Case study 25

Oxfam reported that the emergency relief effort was successful in almost all areas. There were few secondary deaths from starvation, disease epidemics or lack of clean water, even in the isolated Andaman Islands. Children returned to school almost immediately: temporary schools were provided by Unicef for 500 000 children, and Oxfam supplied uniforms. The idea was to try to overcome trauma by returning normality to the children's lives.

However, just as the physical damage between and within countries varied, so too did the recovery. Response rates and the effectiveness of recovery programmes varied. Community resilience depended on the topography, the extent of the damage, the number of deaths, and the timeliness and effectiveness of assistance from various providers. The cohesiveness of the community and its access to social, economic and political resources played a key role in recovery.

In Thailand, for example, there was great success in the fishing community of Kohlanta but much slower progress in Phi Phi. The biggest problem during the emergency period was the breakdown of infrastructure and communications, which meant that organisations lacked transport to reach remote communities.

The rehabilitation phase illustrated the differential progress among communities within the same country and among different countries. Many people reported the almost indecent haste to get Phuket functioning as a premier tourist resort again, at the expense of poor communities in Krabi or Khao Lak. Much depended on how well the sequence of aid was organised locally and how well the providers worked with each other and the local communities.

The Cash for Work programme was a major success. Sixty-four thousand hectares of agricultural land were damaged or contaminated by coastal flooding and over 1 million jobs were lost. Fishing, small-scale subsistence farming, tourism and labouring work were worst affected. The Cash for Work scheme found work for survivors building boats, desalinating land, building and organising village communities as trade cooperatives. Around 60–70% of fishermen are back in business.

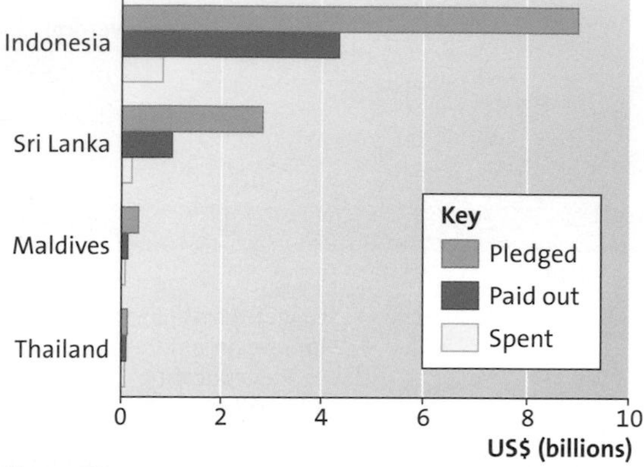

Figure 91
Tsunami aid

However, some major concerns have been raised:

- There was a shortfall between money pledged and money received (Figure 91). Only around 10% of the money pledged had been spent by summer 2006.
- There have been disputes between short-term projects and longer-term plans, for which money needs to be set aside.
- There has been competition between some charities, and emphasis on eye-catching projects such as sponsored fishing boats bearing logos, which proved unsuitable for use by local fishermen.

Progress on re-establishing infrastructure and public services, and rebuilding housing has been slow in comparison with restoration of livelihoods and provision of public health, water and sanitation (Figure 92). The biggest problem relates to land ownership. The authorities in Sri Lanka have planned a buffer zone to prevent fishermen rebuilding houses on the coast. There are also arguments between local people and developers, who have earmarked new land for upmarket hotels as part of Sri Lanka's tourist strategy. In eastern Sri Lanka, rebuilding has been hampered by the re-emergence of the Tamil Tigers, which makes reconstruction both difficult and dangerous.

Figure 92
Tsunami fund spending (US$ millions), January–September 2005

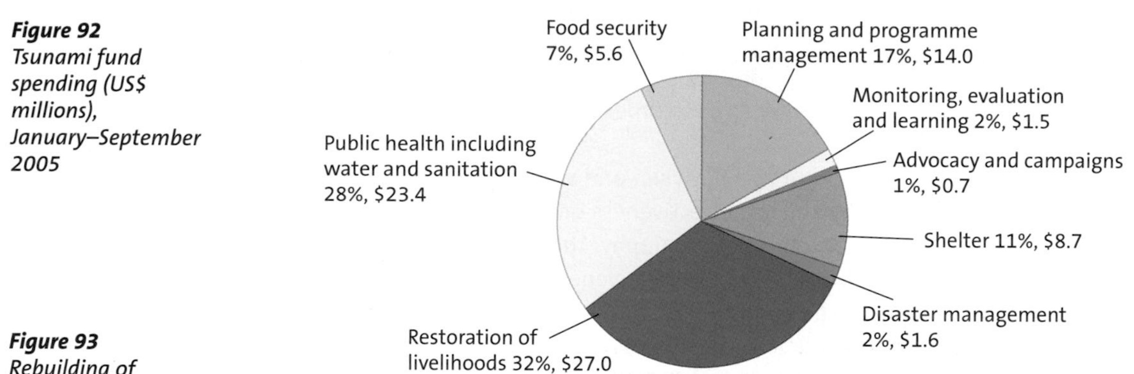

Figure 93
Rebuilding of housing in summer, 2006 after the 2004 tsunami situation

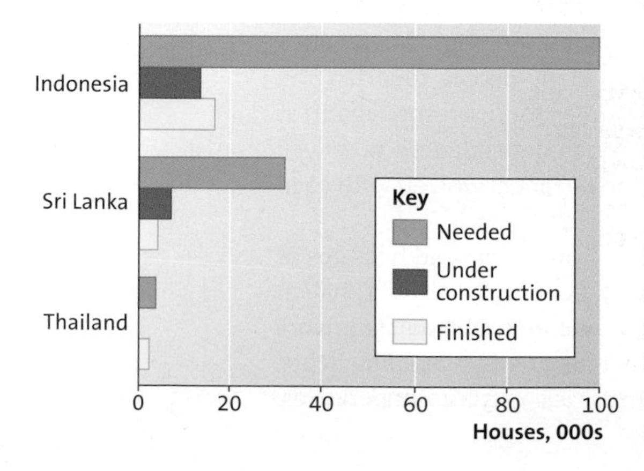

The greatest problems in building new houses are experienced in Sri Lanka and Indonesia (Figure 93). Allegations of corruption centre on the differential progress in various villages. The scale of the disaster (2 million homeless) has meant that many people are housed in legionary barracks (50000 in Indonesia) and transitional houses in Sri Lanka. Decisions on where to rebuild raise issues of personal trauma and memories of lost family members.

The differences in spending between countries is largely proportional to the degree of damage, but there are anomalies based on political circumstances (Figure 94). Myanmar (Burma) has refused to acknowledge the impact of the disaster and has received limited aid from China. The Maldives has repaired the tourist hotel islands but allowed only restricted aid to reach the islands where local people live. India, conscious of its status as an NIC, has provided abundant government aid and donated aid to other poorer countries. However, foreign aid (except for emergencies) has not been permitted in the Indian owned Andaman Islands because of their sensitive military status.

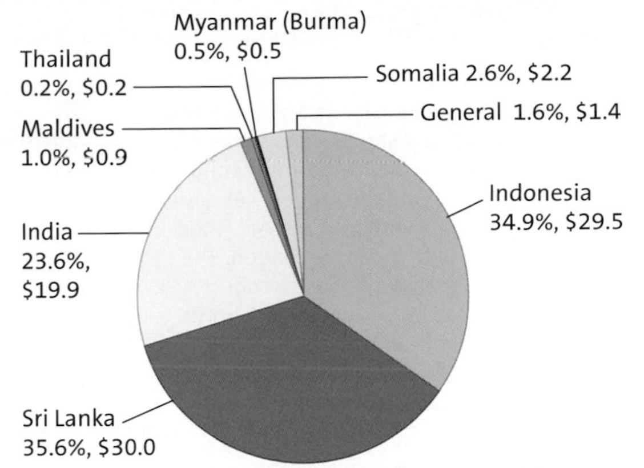

Figure 94 ***Tsunami aid spending (US$ millions) by country in 2006***

Lessons learnt by the end of 2006

A number of issues concerning the disaster management and relief effort have been raised by the UN, the World Bank and Oxfam:

- the need to establish a world disaster fund, contributed to by most nations, so that the UN Disaster Emergency Committee has funding readily available to provide transport and emergency relief
- the need for mitigation strategies to be an integral part of pre-disaster and recovery planning — it is difficult to persuade people that such an event could happen again (World Bank)
- the protection of coastal areas and the location of new buildings away from the coast are therefore key long-term strategies, in spite of local opposition
- the need to establish a legal framework for land tenure, so that disputes about ownership are not a barrier to rehabilitation
- the need to develop procedures to ensure that public works and building can take place, so avoiding reconstruction delay and thousands of people in temporary homes
- to ensure that all projects are equitable with the technology available to the local people, and to empower local people to play a full role in decision-making after recovery, i.e. to ensure sustainability (described by Oxfam as accountability to beneficiaries)

These points need to be supported by the techo-fix of a yearly warning system.

Insurance

Insurance is perceived as a key strategy for MEDCs. Individuals need to realise that the benefits of purchasing an insurance policy will more than outweigh the costs. However, insurers — aware of the huge payouts incurred following major hazards — assess the risk and charge accordingly. In high-risk areas, the premiums are cripplingly high (if insurance is offered at all), so householders are encouraged to take preventive measures, for example earthquake-resistant building materials and flood 'skirts'.

With a potentially exponential rise in global-warming-induced hazards, insurance is becoming a limited option. Hurricane Katrina showed that even in a rich MEDC, many poor people in the city of New Orleans did not have insurance. In some hazard-prone areas, people have inaccurate perceptions of risk and assume the problem will not occur again. However, insurance is useful for commercial organisations who can afford the high premiums.

In conclusion, while modifying the loss is an option, it will never be as significant as modifying vulnerability. *Case study 26* shows how some victims of Hurricane Katrina are still struggling to sort out appropriate insurance payouts and receive government grants 20 months after the disaster.

Case study 26 MODIFYING THE LOSS: AN UPHILL STRUGGLE POST KATRINA

Figure 95 *People profiles post Katrina*

A is a single pensioner. She has moved to Florida and started a new life, supported by her three adult children who are paying the rent on her new apartment. Her house did not flood but it is badly storm-damaged. She went back to her house to collect her belongings but will never return to live there because she does not feel safe. If she gets a grant she will repair the house and try to sell it.

B is a single parent with a teenage son and daughter. They are currently living in a trailer park in Huntingdon, West Virginia. B lost her sister, brother, father and a nephew and cannot face returning to New Orleans. She received an insurance payout after 10 months but the amount is still in dispute. B is unable to work as a result of the trauma. She has $35 000 in credit card debts and these will increase as the children go to college.

C is a single pensioner. His house was underinsured and he is forced to live in his partially rebuilt but still badly damaged house. He lives on a fixed pension of $1 000 per month. He has personal debts of $40 000. He is still awaiting a state grant to rebuild his home.

Mr and Mrs D are a middle-aged couple. Their home was completely destroyed so they moved to their daughter's house in Aberdeen, Missouri. Mr D was a high-school teacher, but the school was destroyed and so he lost his job. He is now working in a supermarket for 30% of his former wage. This couple had hurricane insurance only, no flood insurance, so they are awaiting a state grant. Their mortgage lender gave them a 6-month reprieve, but only the intervention of a local consumer group prevented foreclosure as they are still paying a mortgage on a destroyed house.

	Louisiana	Mississippi
Grant applications	132 778	188 143
Grants paid	9 131	1 260
%	7%	67%
Value of grants paid	US$5686 million	US$882 million
Homes damaged	287 000	–

E is a local business owner. He and his wife boarded up their house and bought and furnished a new house 20 miles away inland. They are currently paying two mortgages and have $50 000 credit card debts as a result of insurance disputes at the former home and business, which were both affected by flooding. If a settlement is not made, E is considering filing for bankruptcy. He has applied for a state business-regeneration grant.

F is a single 85-year-old woman. She lives beside her old house in a trailer supplied by FEMA. She received $65 000 from her flood insurance policy, but this was not enough to rebuild. She has heard that she will get a state grant, perhaps in 6 months time.

19 Question

Using case studies

(a) What are the main problems that people are experiencing when trying to modify the loss from Hurricane Katrina?

(b) What sort of profile do the most seriously affected people have?

Guidance

(a)
- Explore insurance and grant-aid issues.
- Think about the amount of damage.
- Consider issues of high cost of migration and lack of economic recovery.

(b) Consider the nature of the group of people, largely pensioners, or those with limited incomes.

Part 8

Global trends in the occurrence and impact of hazards

Figure 96 summarises four key trends in natural disasters. It is based on statistics gathered by the Centre for Research on the Epidemiology of Disasters (CRED) between 1970 and 2000.

- The number of people reported killed dropped dramatically after the 1970s and then levelled off — the statistics for 1980–89 and 1990–99 are similar.
- The number of people reported affected by hazards/disasters (e.g. injured or with loss of livelihood) has increased each decade.
- The economic losses have grown exponentially.
- The number of reported disasters has grown significantly by about 800 each decade.

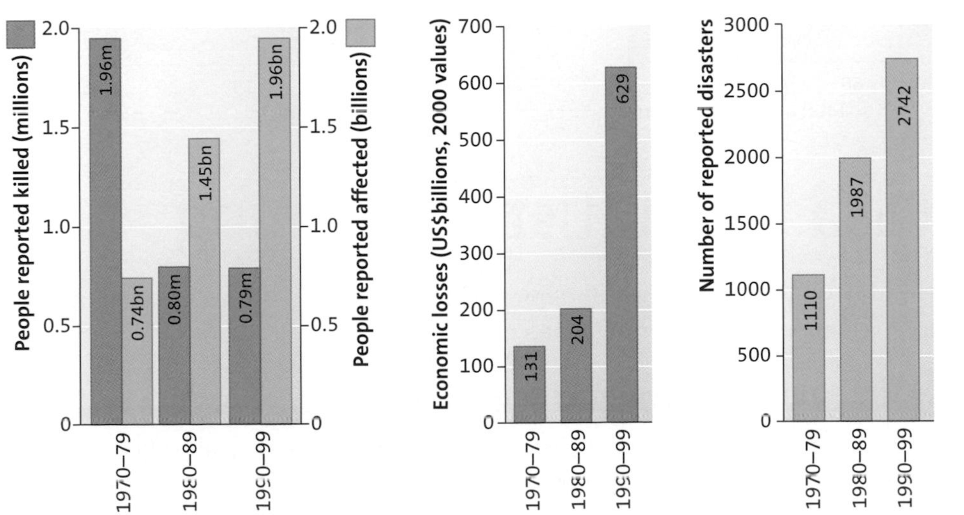

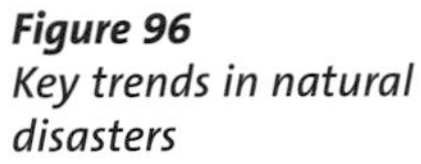

Figure 96
Key trends in natural disasters

However, the trend in recent years is influenced by anomalous years. Natural disasters made 2005, for example, an unforgettable year. It began with the aftermath of the Asian tsunami, which killed around 300 000 people and left 1.5 million people homeless. It was a record hurricane season, including Katrina, Rita and Wilma. In October, the earthquake in Kashmir claimed nearly 75 000 lives. An estimated 150 million people were affected by 360 disasters. A US$160 billion bill for damage (economic costs, 80% of which was for Hurricane Katrina) represented a 71% increase from 2004.

Figure 97 compares the human impact of natural disasters for 2001–05 with 2006. Apart from windstorms, 2006 was a relatively quiet year compared with the average for previous years.

Figure 97* *Human impact by disaster type

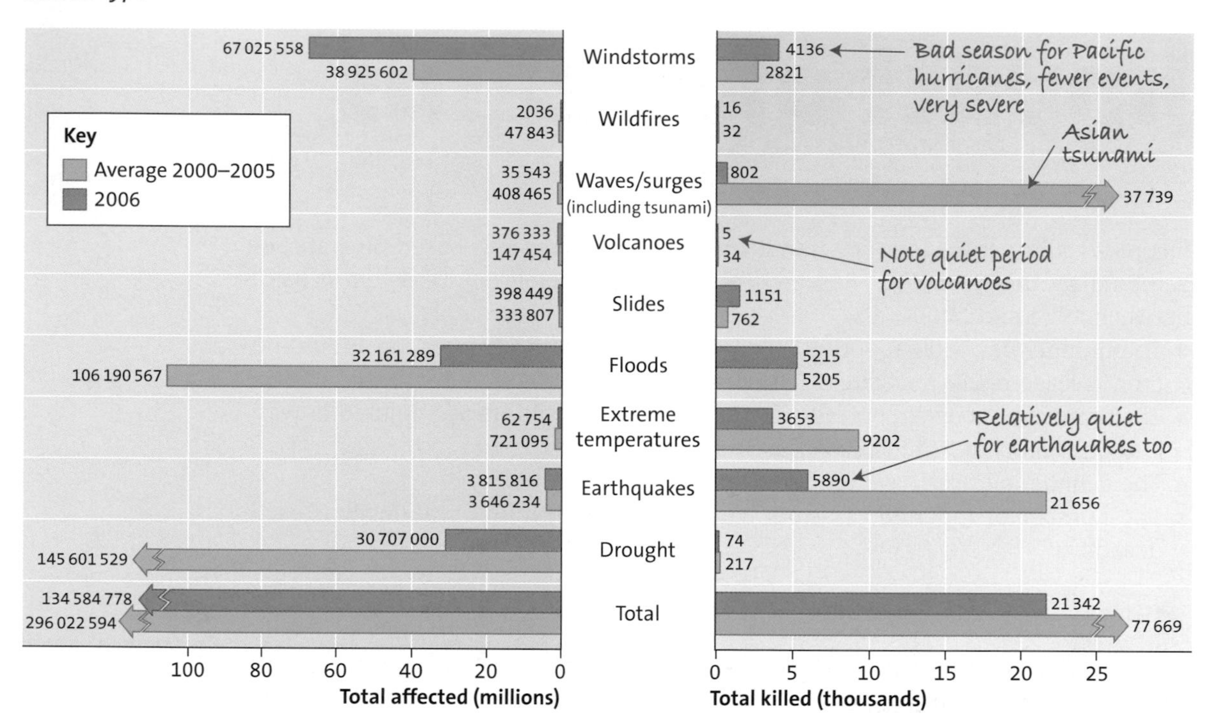

Disaster statistics warning

Disaster statistics need to be interpreted with great care and you need to adopt a forensic approach to interpretation.

Is the source of the data reliable?

Governments report disaster statistics to UN agencies. These quasi-official statistics are supplemented and cross-checked by designated groups that also monitor reports from the media (internet data are important) and from NGOs, such as Oxfam, working at the 'front line'. The pre-eminent authority is CRED, supported by the World Bank and insurance companies such as Munich Re, Swiss Re and Lloyds of London.

How good are disaster statistics?

There is neither a universally agreed definition of a disaster nor a universally agreed numerical threshold for disaster designation. Reporting disaster deaths is controversial, because it depends on whether direct (primary) deaths or indirect (secondary) deaths from subsequent hazards or associated diseases are counted. Location is significant because local or regional events in remote places are under-recorded.

Declaration of disaster deaths and casualties may be subject to political bias. The 2004 Asian tsunami was almost completely ignored in Myanmar, but perhaps initially overstated in parts of Thailand, where foreign tourists were killed, and then played down to protect the Thai tourist industry.

Statistics on major disasters are difficult to collect, particularly in remote rural areas of LEDCs (e.g. the earthquake in Kashmir in 2005) or in densely populated squatter settlements (e.g. the Caracas landslides, 2003–04).

Time-trend analysis (interpreting historical data to produce trends — see Figure 96) is difficult. Much depends on the intervals selected and whether the means of data collection have remained constant. Trends can be upset by a cluster of mega-disasters, as happened in 2005–06.

Number of deaths from disasters

Globally, the annual number of deaths from disasters seems to have fallen, or at least levelled off (Figure 96). Ninety percent of such deaths occur in developing countries. Twenty-four of the 49 least developed countries (LDCs) face high levels of disaster risk, with six countries experiencing between two and eight major disasters during each of the last 15 years.

The decline in disaster-related deaths should be set against the backdrop of rising hazard numbers and increasing numbers of vulnerable people. The **risk disc** (Figure 98) explains the reasons for the decline in deaths in terms of disaster preparedness, disaster mitigation (hazard proofing), disaster response and disaster recovery.

In 1994, in Yokohama, at the mid-decade conference for the International Decade for Natural Disaster Reduction, an important watershed was reached. Delegates argued for a less **top-down** technocratic fix and a more **bottom-up** approach, using NGOs to mobilise local communities (particularly in large cities) in strategies to improve their resilience to disaster. It is the combination of these two approaches (techno-fix and community preparedness) that has led to the reduction in deaths.

Figure 98
The risk disc

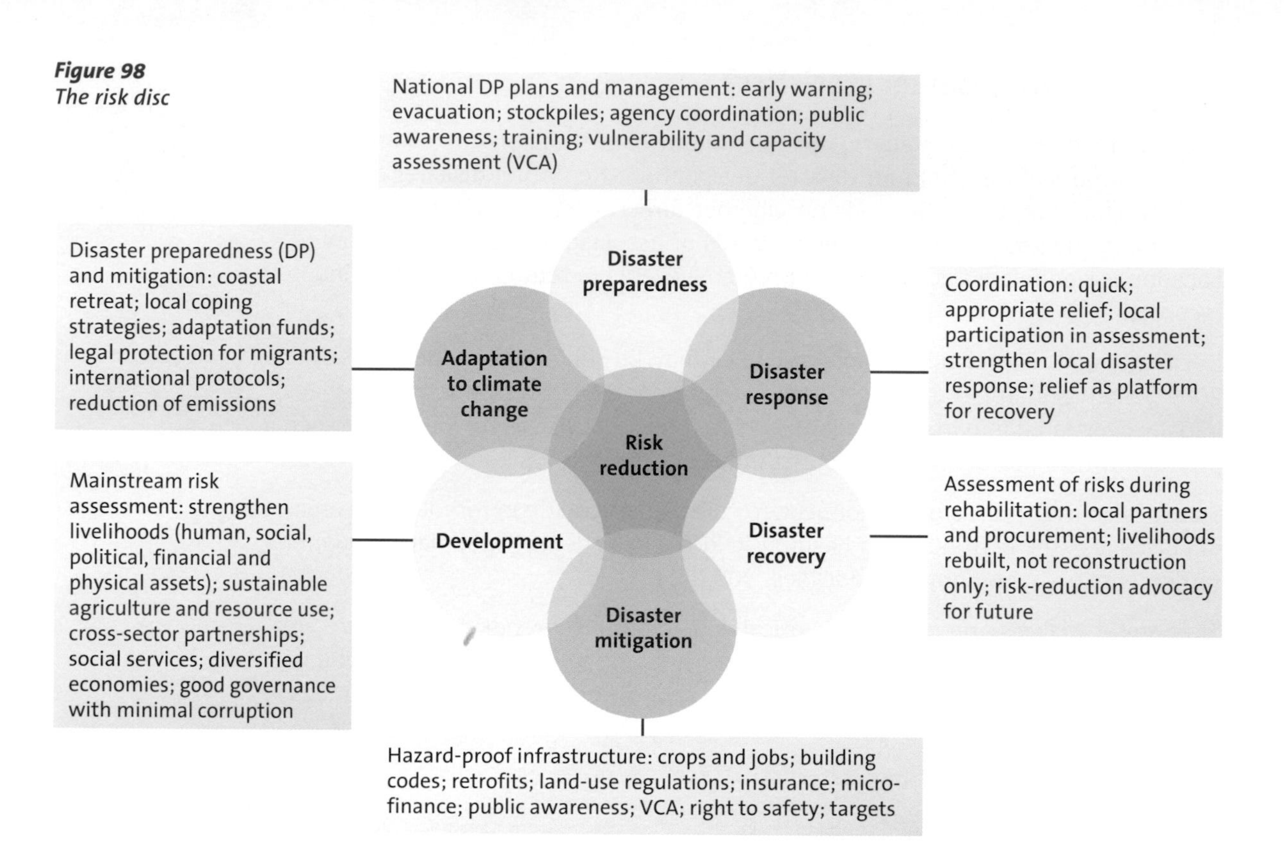

Responses to the savage storms of 2004

In 2004, a series of hurricanes crisscrossed the Caribbean. The responses of countries were largely dependent on the level of technology and the degree of political stability.

In Cuba, the response successfully combined the political efficiency of a totalitarian state with sound technology and outstanding community preparedness. Civil defence is now part of national security and during Hurricanes Charley and Ivan only four people were killed. Cuba has a world-class meteorological institute that produces excellent computer models and reaches people through all types of media.

There is education on the dangers of hurricanes and the systems are well-consolidated at national, regional and local government levels.

In Jamaica, preparedness paid off. Since Hurricane Gilbert, there have been community disaster response officers and plans in each parish. Factors include:

- good maps showing resources and the homes of the most vulnerable people
- trained and well-equipped Red Cross emergency helpers, wardens, drugs, refuge shelters (e.g. all schools)
- voluntary annual Gilbert anniversary practice drills

Fourteen people were killed in Hurricane Ivan.

In the Dominican Republic, the focus was on wind hazard, yet the main problem in 2004 was flooding (100 mm in 24 hours). Positive factors now in place include:

- delivery of maps, and visits from Red Cross outreach workers
- a new civil defence system based in churches

A problem was a lack of short-wave radios and other key equipment. Twenty people drowned during Hurricane Jeanne.

Haiti is the poorest Caribbean country. A political vacuum led to lack of local government organisation. Extreme poverty has caused a lack of resources — for example, at the national meteorological centre there is a lack of local systems. The people are poorly educated, so it is difficult to develop communications effectively.

During Hurricane Jeanne, up to 3000 people died from mudslides caused by torrential rain.

Number of people affected by hazards and disasters

The number of people affected annually by hazards and disasters has increased considerably (Figure 96). 'Being affected' includes loss of home, crops, animals or livelihoods, or decline in health for a designated period of time (often 1–3 months). On average, 200 million people per year are affected by disasters.

There are a number of interlinking factors that have led to the increasing number of people, particularly in LDCs, being vulnerable to disaster (Table 17).

***Table 17** Factors leading to increased vulnerability*

Factor	Impact on vulnerability	Example
Population growth and change	The world's population should stabilise at 9 billion in 2050; the growing proportion of the very young and very old will increase the number of people affected by disasters	Growing concern over the vulnerable aged in Japan and Florida, USA
Land pressure	Land pressure leads to the occupancy of areas of high risk, the cutting down of forests to provide farmland and the destruction of mangroves to allow coastal development	In Bangladesh, 85% of the population depends on subsistence farming; millions live on disaster-prone floodplains
Urbanisation	The rapid rate of urbanisation in the developing world is a key factor in the growth of vulnerability, particularly of low-income families living in squatter settlements/slums; squatter settlements occur in zones of high risk; slums grow up in inner-city tenements; many megacities are in multiple-hazard zones	Caracas landslides, 1999; Mexico City earthquake, 1985, which demolished the slum areas; the garbage dump mudslides of Manila City
Political change	This can lead to destabilisation; corruption or war can exert extra land pressure and lead to desperate responses by rural migrants	Darfur, Sudan; Haiti — between governments during Hurricane Jeanne
Economic growth	This leads to development of sophisticated structures such as bridges; it can also lead to loss of forests, soil degradation and overuse of resources, all of which increase vulnerability	Texas and Atlantic coasts in the USA are vulnerable to hurricane damage, e.g. bridges that link barrier islands
Globalisation	This has led to TNCs being major world players who exploit natural resources	Timber companies in the Philippines; tropical storm Debbie led to major flash flooding and siltation in St Lucia, where watersheds had been deforested for banana plantations

Factor	Impact on vulnerability	Example
Technological innovation	Flood barriers have been built along rivers and coasts but it is usually too costly to prepare for even a 1-in-200-year event; false security may encourage building in unsafe areas	In New Orleans, the river flood-control banks had not been strengthened to cope with a disaster of Hurricane Katrina's proportions
Development gap and the occurrence of poverty	As the development gap widens between LDCs and the rest of the world, the problem of debt servicing means that many countries have no money available for disaster management	DFID is now providing grants (e.g. to Bangladesh) specifically for flood action and cyclone action plans
Climate change	Human-induced global warming is causing long-term climate change that could lead to deforestation or desertification; short-term extreme weather events such as droughts/fires and floods occur; increased ocean warming may be spawning more hurricanes	The 1989 and 1997 El Niño events were felt around the world; European floods 1999–2000, 2002, and European droughts in 2003 and 2006 could be related to climate change

Economic losses from disasters

Economic losses from disasters have grown exponentially (Figure 96). Comparing 1980–89 and 1990–99 shows that economic losses tripled. Losses are growing at a far greater rate than the number of disasters.

Munich Re, the reinsurance giant, looked at the trend of economic losses and insurance costs over a 50-year period (Figure 99).

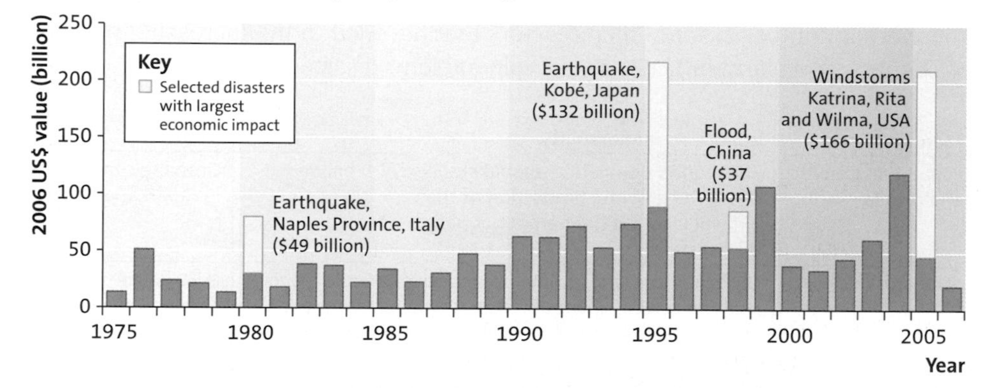

Figure 99
Trends in economic losses, 1975–2006

Rising trends can be identified. 2006 was an anomalously quiet year, but mega-disasters such as Hurricanes Andrew and Katrina lead to huge rises for a single year, resulting in a fluctuating trend. The trend of insured losses is rising less rapidly than that for total economic losses. This is a reflection of:

- the spatial concentration of hazards
- their realisation into disasters in LEDCs
- the poverty-stricken underclasses in MEDCs not being able to afford the high cost of insurance in hazard-prone zones (e.g. some districts of New Orleans, see *Case study 26*, p. 94).

There is a tendency to over-emphasise the economic damage and losses experienced in MEDCs. In absolute terms, there is no doubt that because of the value of their economies, the sophistication of their installations and infrastructure, and the scale of domestic insurance claims, the amounts involved will be large. Figure 100(a) compares insurance losses of recent mega-events. In human costs the rank order is different (Figure 100(b)), reflecting the MEDC/LEDC divide. This is often expressed by the simplistic statement 'MEDCs experience the greatest damage (economic cost), whereas LEDCs experience the greatest number of deaths (social cost)'.

***Figure 100** (a) Insurance losses and (b) human costs of recent mega-events*

(a) Insurance losses ($bn) 0 2 4 6 8 10

Event	Insurance losses ($bn)	No. of victims dead or missing
USA/Bahamas *Hurricane Katrina*	45	1326
USA/Cuba *Hurricane Rita*		34
USA/Caribbean *Hurricane Wilma*		35
Europe *Winter Storm Erwin*		18
Europe *Rain, floods and landslides*		49
USA/Caribbean *Hurricane Dennis*		65
India *Floods and landslides*		1150
USA *Thunderstorms*		n.a.
UK/Ireland *Storm Gero*		9
Asia *Typhoon Nabi*		34

(b) Number of victims dead or missing 0 500 1000 1500 2000

Event	Number of victims dead or missing
Pakistan/India *Earthquake*	73 300
Pakistan *Cold weather*	
Latin America *Hurricane Stan*	
USA/Bahamas *Hurricane Katrina*	
Indonesia *Earthquake*	
India *Floods and landslides*	
Iraq *Stampede*	
Iran *Earthquake*	
India/Bangladesh/Nepal *Heatwave*	
India *Stampede*	

In reality, the situation is economically more complex. In *relative* terms, economic damage from natural disasters tends to be highest in LEDCs, mainly because of their high dependency on one or two cash crops or on tourism. The damage forms a high percentage of their annual GDP (Figure 101).

Disaster	Percentage of GDP
Nicaragua (Hurricane Mitch, 1998)	50.0%
Honduras (Hurricane Mitch, 1998)	37.7%
Tonga (cyclone, 2001)	36.2%
Belize (hurricane, 2000)	34.4%
Yemen (flood, 1996)	20.7%
El Salvador (earthquake, 2001)	20.0%
Jamaica (flood, 2002)	14.0%
Indonesia (forest fires, 1997)	7.9%
Papua New Guinea (volcano, 1994)	7.5%

***Figure 101** Percentage of GDP represented by losses from some natural disasters, 1994–2002*

MEDCs have more diverse economies that are better able to withstand natural disasters. For example, Munich Re estimated that between 1994 and 2003, Japan suffered US$166 billion of economic damage from natural disasters (including the Kobe earthquake). This was only 2.6% of its GDP (compare the LEDCs in Figure 101).

In conclusion, the rate of economic losses is increasing much faster than the occurrence of disasters, largely because of the growing economies of many RICs and NICs, especially in Asia.

Number of reported disasters

The number of reported disasters has grown by about 800 for each decade (see Figure 96). When profiling disasters, key features are the frequency of occurrence and the magnitude of the event. The trends in frequency and magnitude affect their impact in terms of economic and social costs.

Figure 102 looks at the rising trend by major disaster type. Earthquakes, storms (including hurricanes) and floods comprise around two-thirds of all major disasters. In recent years, drought has become more widespread and significant, affecting millions of people (10%). Other disasters, such as volcanic eruptions (2%), tsunamis (1%) and avalanches (1%) are 'rare', but can be devastating. Figure 102 shows that

Figure 102
(a) Number of natural disasters, 1900–2005; (b) hydrometeorological disasters, 1970–2005; (c) geological disasters, 1970–2005

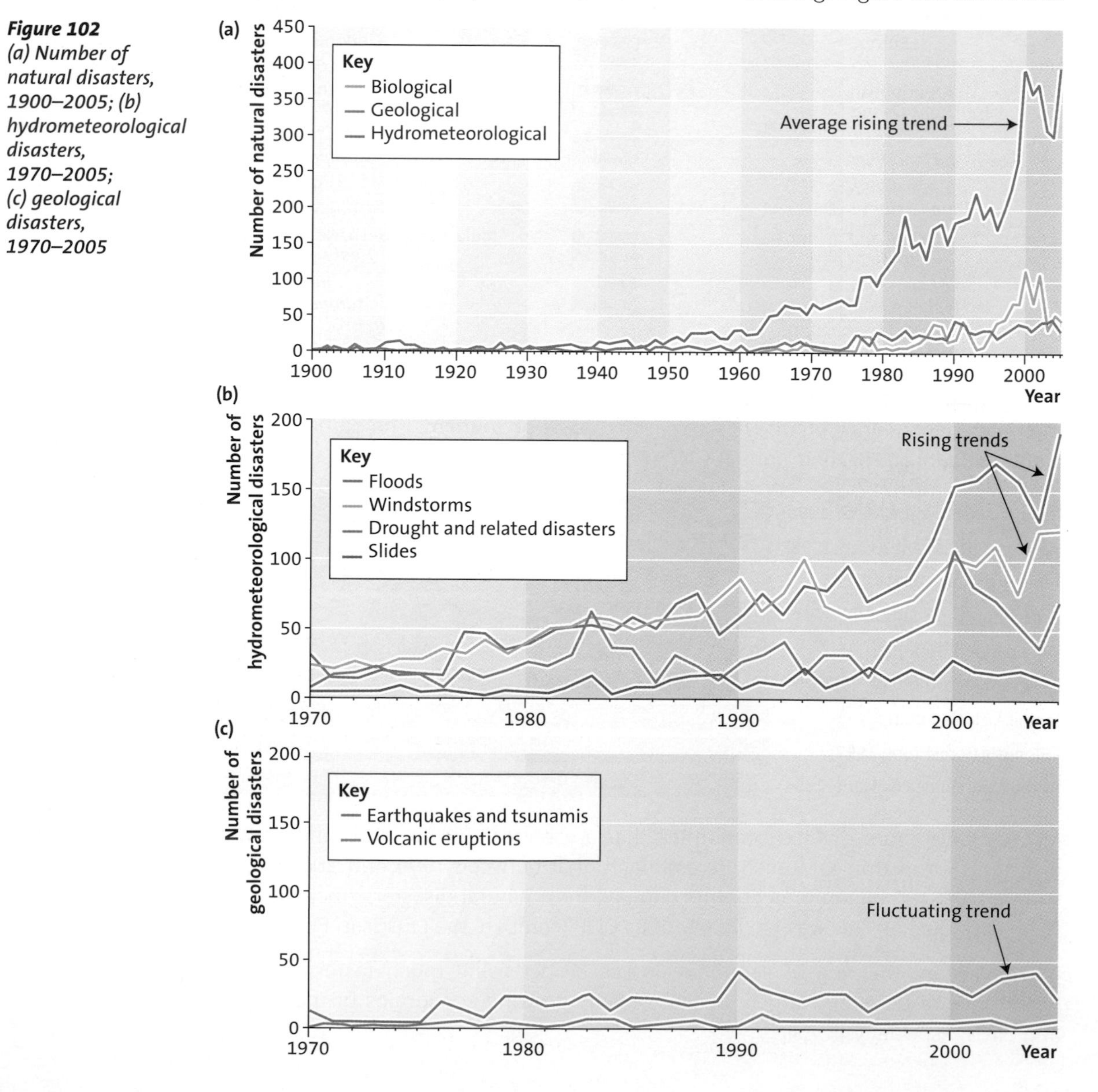

the rising trend is a result of hydrometeorological hazards (floods, storms and droughts); earthquakes show fluctuations (long timescale variations) but no rising trend.

Consider the incidence of tectonic hazards. Timescale variations can be explained by spatial clustering along mobile plate boundaries. Earthquakes, and their associated volcanic activity, frequently occur in series, such as those occurring along the Sunda fault off the coast of Sumatra, where the Indian plate is subducted beneath the Burma plate. Scientists liken the Sunda fault to a zip that is gradually coming undone as each earthquake transfers stress along its length. Recently there have been two major earthquakes in Sumatra (2005) and one in Java (2006), with two tsunamis generated as a sudden change occurred in the sea floor. Further activity occurred in both Java and Sumatra throughout 2007.

Arguments put forward to explain the increase in the number of reported hydro meteorological events are usually associated with climate change. Table 18 shows ways in which the Intergovernmental Panel on Climate Change predicts how a more extreme climate (resulting from global warming) will have a hazardous impact on people and environments.

***Table 18** Examples of possible impacts of climate change due to extreme weather fluctuations and climatic events*

Projected changes during the twenty-first century	Possible impacts
Higher minimum temperature; fewer cold days, frost days and 'cold waves' over nearly all land areas (certain)	• Decreased deaths from hypothermia • Changing patterns of crop growth • Extended range and activity of some pest and disease vectors • Reduced demand for heating energy • Changing winter tourism patterns
Higher maximum temperature; heatwave over nearly all land areas (very likely)	• Increased deaths of elderly and urban poor (Paris, 2003) • Increased heat stress of livestock/wildlife • Shift in tourist destinations • Huge demand for air-conditioning (failure of electricity supply in California, 2006)
More intense precipitation events (storms) over many areas (very likely)	• Increased flood, landslide, avalanche, mudslide and debris-flow damage • Increased soil erosion • Increased flood runoff, so more pressure on government, private flood insurance and disaster relief schemes
Increased summer drying and associated risk of drought in most mid-latitude continental interiors (likely)	• Increased risk to quantity and quality of water resources • Increased risk of forest fires • Increased subsidence/shrinkage of building foundations, especially in clay areas • Decreased crop yields
Increased tropical cyclone peak wind intensity, mean and peak precipitation intensity (likely)	• Increased risk to human life, and of infectious diseases and epidemics • Increased coastal erosion and damage to coastal buildings and infrastructure • Increased damage to coastal ecosystems (e.g. corals and mangroves)
Intensified drought and floods associated with the El Niño–La Niña cycle, with tele-connections across many regions (likely)	• Decreased agricultural and range-land productivity (e.g. midwest USA, flood and drought prone) • Decreased HEP potential in some drought-prone regions (e.g. New Zealand)
Increased variability of Asian monsoon precipitation (likely)	• Increased flood and drought magnitude and damages in both temperate and tropical Asia (India, 2006)
Increased intensity of mid-latitude storms (likely)	• Increased risks to human life and health, especially on the coast • Increased property and infrastructure losses

Projected climate change in coming decades is likely to alter the frequency and magnitude (intensity) of climate hazards:

- Changes in precipitation patterns, soil moisture and vegetation cover could lead to more frequent droughts, floods and subsequent landslides.

- Increased temperatures will warm the oceans, which could spawn more intense storms (category 4 or 5). In spite of much discussion about the frequency of hurricanes being related to global warming, most scientists link this more to oscillations such as the El Niño southern oscillation. They note that hurricanes are minimal in the Atlantic in El Niño years, but at their worst in La Niña years. Current research suggests that seasonal oscillations between the Atlantic and Pacific influence the occurrence of storms.
- Global warming is leading to thermal expansion of the oceans. Rising sea levels are creating higher storm surges during hurricanes (Hurricane Katrina).
- Global warming is leading to more unpredictable climates with more extreme weather events such as the European floods in 2002 and European droughts in the summers of 2003 and 2006.
- Some scientists suggest that global warming is leading to more frequent and intense El Niño events, with 1997–98 being the strongest ever recorded. The El Niño–La Niña cycle led to numerous hydroclimatic hazard events (e.g. floods in Peru and drought/fires in Australia, 2006–07).

Another explanation for increased frequency of hydroclimatic disasters lies with increased environmental degradation caused by population pressure, which leads to deforestation and other land-cover loss. This can lead to flash flooding. Scientists are divided as to whether watershed deforestation and removal of land cover is the root cause of increased flash flooding in, for example, China, Vietnam and Bangladesh. Environmental degradation influences the effects of natural hazards by exacerbating their impact.

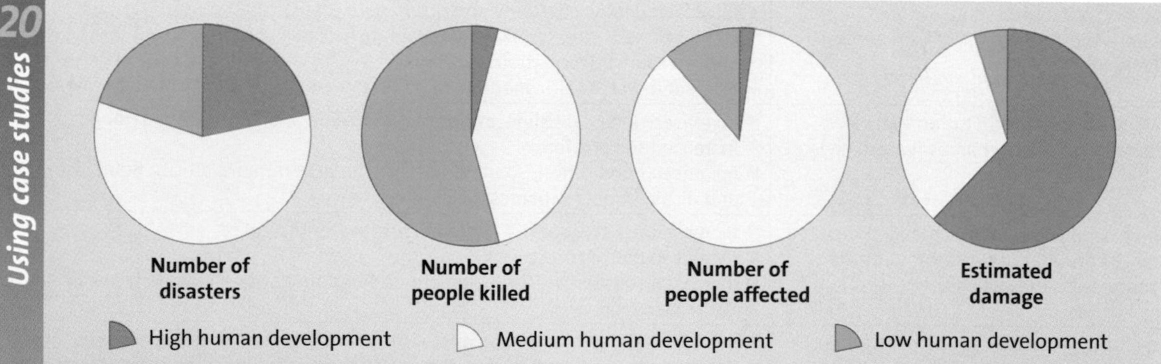

***Figure 103** Disasters related to human development*

(a) Describe the variations in spatial impact of these global trends and suggest reasons for the variations.

(b) Assess the implications of these trends.

Guidance

- The number of hydrometeorological disasters is expected to increase, particularly in Asia, Europe and the Americas. This will affect many people in NICs/RICs and middle-income countries. The number of deaths continues to rise in LDCs.
- The numbers killed in MEDCs remains low.

Part 9

Advice to students

How to research hazards effectively

Information about hazards is available from a variety of sources. The information in newspapers is often text-rich with occasional photographs. In more specialist publications, such as *New Scientist, Scientific American* and the *Economist*, information is presented in more technical language, with diagrams. Valuable information can come from listening and watching. Radio programmes (particularly those on BBC Radio 4) can be a good source of documentary evidence. Digital satellite channels such as Discovery and BBC4 can provide valuable, up-to-date information.

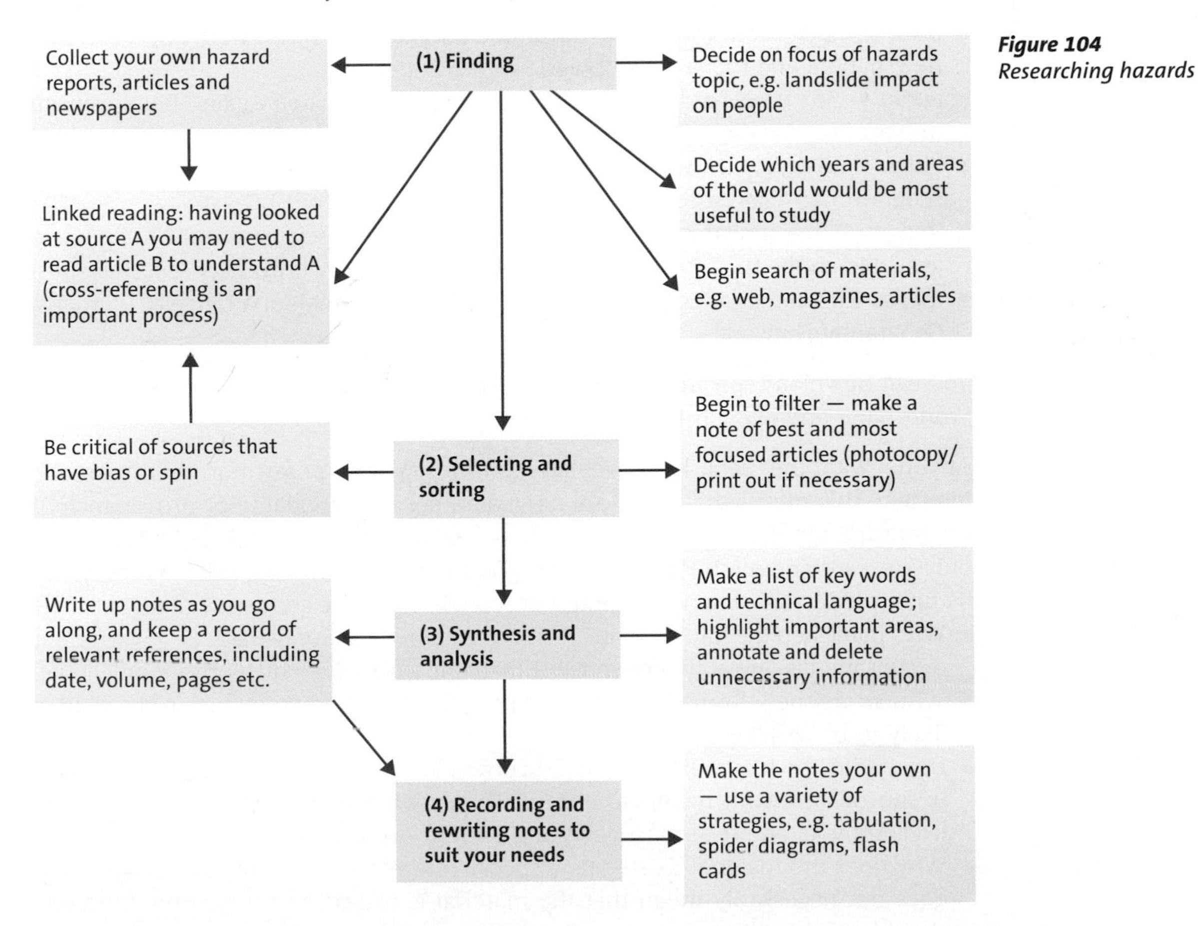

Figure 104
Researching hazards

Magazines and newspapers

Table 19 gives details of selected publications and their relevance to hazards research.

Table 19
Printed resources

Publication	Description
Economist	A weekly publication with short, accessible articles relating to hazards
New Scientist	A weekly publication with accessible style and language, high-quality diagrams and supporting illustrations; good supporting website
Scientific American	A monthly magazine with some interesting articles; written in a style that tends towards the academic
Geography Review	Published 4 times per year; reports are concise and factually strong (search the index at the end of each volume for relevant hazards articles)
Guardian, Independent, Daily Telegraph	These newspapers often report on hazard-related information; good summaries of events a couple of days post-impact

Using the internet effectively

There is a mass of information on the internet (see *Websites*, p. ix), such as photographs, blogs, documents, video, audio (including RSS news feeds), articles, diaries and maps:

- print — pdf files, Microsoft Word documents and html web pages, online newspapers and magazines, 'blogs'
- audio — RSS news feeds, specialised podcasts, BBC's 'Listen Again', Blinx website (to search for audio feeds)
- video — CNN, ITN News, British Pathé (archive). You can watch some broadcast programmes again through the internet, e.g. Channel 4, BBC.
- images — use a search engine to find image files. Specialist hazard images are available on the US Geological Survey website and many university research sites.
- maps — Multi-map, Google local, Google Earth, 'Topozone' for a range of US maps, Geography Network

You can download specific reports from the websites of large agencies such as the World Bank and Munich Re.

Anyone with the right hardware and know-how can post information on the internet. This means that there are no guarantees on the quality or provenance of the information. Information services may charge or be accessible by subscription only. Researching on the internet gives you the feeling of being busy but may not be time-productive. When using internet sources, there are some points to consider:

- Who published the information? A site maintained by a university or government organisation is probably more reliable than one maintained by a private individual.
- Who wrote the information? Material provided by a known expert in the field is likely to be reliable.
- How old is the material? If you need current statistics, check the age of the data. A site dealing with historical hazards information may not need updating as frequently as one related to news and current events.
- Why does the material exist? Many special interest groups have web pages. This does not necessarily mean that the material is biased, but it is something you should consider.

Organising notes

To get to grips with ideas and 'make them your own', there is really no alternative to unpacking, distilling and synthesising information from the documents you have collected. This can be a real hatchet job, involving crossing out, cutting off and throwing out parts of the writing. You will need to decide what is particularly important, what is moderately important and what you can do without.

Some articles have useful diagrams, charts and graphs. When interpreting these:

- take time to understand what the chart or graph is telling you
- start by looking at the title, scales and axes; identify units
- select one or two points and make sure that they make sense
- look at the overall shape of the curve(s) — is it linear, exponential, bell-shaped, skewed?
- look for obvious patterns, peaks, troughs, blips and anomalies — what might be causing them?
- examine the source of the data and think about how the information is presented; think carefully about validity before drawing conclusions

Spider diagrams and mind maps are useful as revision tools and can be used to summarise large amounts of text. Figure 105 shows a case study of the eruption of Mount Pinatubo in the Philippines in 1991.

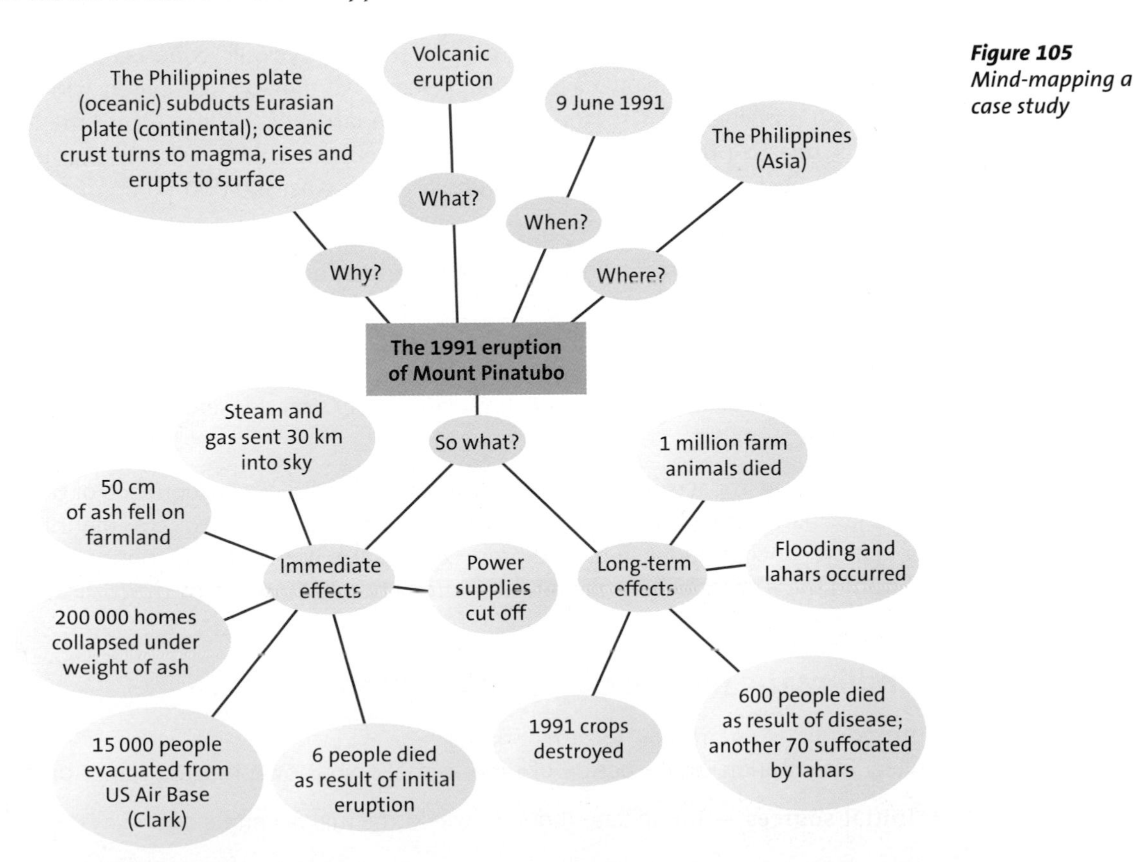

Figure 105
Mind-mapping a case study

Lessons learnt from recent mega-disasters

A mega-disaster is a high-impact event with large-scale loss of life and damage.

Mega-disasters attract widespread (global) media coverage but they need not have a global impact (Figure 106). The 2004 Asian tsunami and Hurricane Katrina (2005) were both mega-disasters and yet they differed considerably in terms of loss of life (300 000 versus 1800) and geographical scale and extent. The primary physical effects of the tsunami were recorded at locations around the globe, whereas Katrina had a relatively limited spread of damage, the primary effects being felt only in New Orleans and surrounding areas.

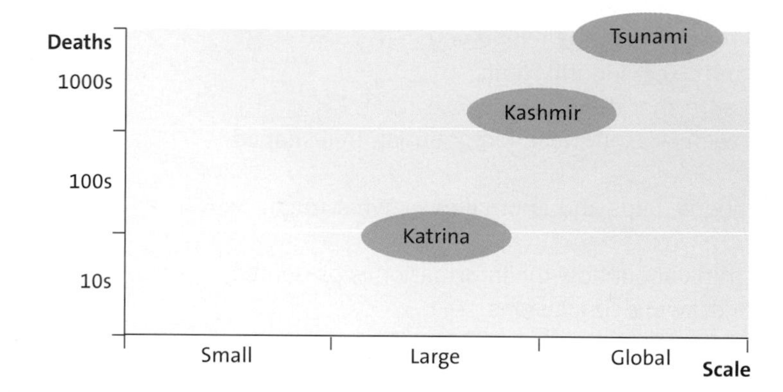

Figure 106
What makes a mega-disaster?

Mega-disasters are likely to be devastating for the particular location. For example, hurricanes and cyclones are a regular feature of the area between 5° and 20° north and south of the equator. So for a tropical cyclone to be considered a mega-disaster, it must involve a significant degree of damage compared with the norm for a particular year/season and probably a large loss of life. Again, the loss of life is relative. Hurricanes Mitch (1998) and Bhola (1970) were both mega-disasters, but the loss of life from the Bhola storm was 500 000 compared with 11 000 for Mitch.

Level of development plays an important role in terms of the relative degree of disaster and scale of impact. The Bhola cyclone was accompanied by high rainfall and a powerful storm surge. Bhola is a low-lying, densely populated area and the community was unprepared. Mitch involved the loss of many lives because its course was unpredictable and erratic. It moved slowly over densely populated areas that were ill-prepared.

The devastating impacts of these mega-disasters are a direct consequence of development characterised by urban growth (often without adequate planning), population concentrations in hazard-prone zones, and human interactions with the environment. These factors magnify the intensity and impacts of mega-disasters in disaster-prone areas.

Handling the mega-disaster: examination advice

Sources

Collect information over a period of time to make your own mega-disaster diary:

- **Initial sources** — immediate (hours/1 week), factual accounts and views

- newspapers
- news websites
- television and radio
- internet blogs

- **Quick articles** — 1–2 weeks, factual reports, initial impacts
  - *New Scientist*
  - *Economist*
  - *Time*
  - *GeoNews*
  - *Newsweek*
  - weekend newspapers
- **Specialist magazines** — 6 months, digest of causes, players and impacts
  - *Geofile Online*
  - *GeoFactsheet* (Curriculum Press)
  - *Geography Review*
  - *Geographical*
- **NGO reports** — 1 year, detailed accounts and response
  - World Bank
  - Oxfam
  - International Red Cross
  - Munich Re
- **Books and journals** — 2+ years, reflective summaries
  - various textbooks
  - *Area*
  - *Transactions of the Institute of British Geographers (IBG)*
  - *Teaching Geography*

Issues

- Too much information — careful selection is required
- Bias and controversy
- Real facts and figures (particularly economic and human costs) take time to emerge
- Mega-disaster impacts are complicated by varying consequences in different countries (levels of development?)

Gearing-up for the exam: example

The 2004 Asian tsunami — why was it a mega-disaster?

- Huge natural event — earthquake 9.1 on the Richter scale
- High vulnerability — people were largely unsuspecting
- High risk — a lot of people were living on the coast
- Physical geography of the coastlines
- Low frequency of tsunamis in the Indian Ocean
- Development factors
- Sheer scale of the impact
- No early warning system

21 **Complete the following table for the 2004 Asian tsunami.**

Basic hazard event profile: magnitude, speed of onset, duration, areal extent, spatial predictability and frequency. Basic facts and figures — dates, times etc.	
Causes: some hazards are entirely natural in origin, such as earthquakes and volcanic activity; causes may be amplified or lessened by human activity, e.g. flooding. Primary versus root causes?	■ Earthquake activity ■ Location ■ Size of tsunami is directly related to (i) the rate of water displacement, (ii) the sense of motion of the ocean floor at the source of the disturbance, (iii) the shape and amount of displacement at the rupture zone, (iv) the depth of the water in the source area.
Spatial impact: importance of distance decay effect. How wide-spread (impacts on different levels of development?)	■ The waves took only 15–20 minutes to reach the coast of Indonesia, and arrived in India zand Sri Lanka within 2 hours. ■ The largest waves travel perpendicular to the alignment of the fault that has generated them. In this case, since the fault is aligned in a generally north–south direction, the largest waves travelled eastwards towards Indonesia and Thailand and westwards towards Sri Lanka, India, the Maldives and east Africa. ■ Locations near to the epicentre, e.g. Aceh (Sumatra), had no time to respond or react since the wave arrived only minutes after the main earthquake. Buildings and infrastructure were also damaged by the earthquake itself.
Analysis of responses: identification of different players and agencies. Spectrum of responses (community versus hi-tech). Management — relate to Parks model.	
Trends: what is the global pattern, e.g. increasing incidence of droughts and floods? Worries about hurricane intensity (not necessarily frequency) and concentration of tectonic activity in southeast Asia. Much controversy surrounding ideas, patterns and trends.	■ The Pacific Ocean comprises over 75% of the world's active and dormant volcanoes, as well as being home to the majority of the Earth's undersea subductive earthquake fault lines. ■ This tectonically active area is probably primed for another 'zip effect' event. ■ There may be a longer term pattern to the events that happened here. ■ The most developed tsunami warning systems are located in the Pacific Ocean.

Drawing useful maps and diagrams

You need to carry out a personal cost–benefit analysis. The main 'cost' is the time taken to draw the diagram or map, but the benefits can be enormous if it is well designed.

Useful diagrams include:

- simple flow diagrams to show a sequence of events
- diagrams to show complex processes such as plate movements or hurricane formation
- comparative tables that summarise information
- diagrams of key models, such as the 'response curve'

You have to balance simplicity and clarity against accuracy and detail.

Diagrams

Figure 107 shows a block diagram of a destructive plate boundary. Unless you are a good artist, you might struggle to draw a block diagram in a short time. Draw just the front of the block to provide a cross-section.

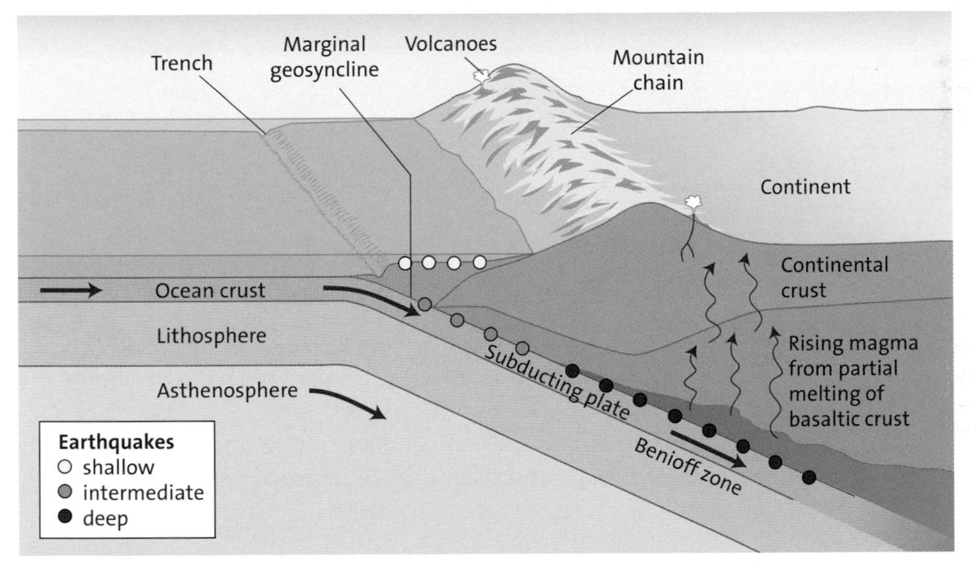

Figure 107 *Cross-section of an oceanic/continental destructive margin*

Select annotations that refer to your arguments — the process of **subductional** melting, rising magma, andesitic volcano formation; the range of earthquake types associated with the **Benioff zone** could also be relevant. The differential density of the oceanic and continental crust is also important.

You could customise your diagram to a particular location by using the associated text:

> The oceanic **Nazca** plate is moving east approximately 12 cm per year and converges with and **subducts** beneath the continental **South American plate**, which is moving west. **The Andes**, a chain of fold mountains rising to nearly 7000 m above sea level, has been formed as the continental crust has been buckled and uplifted. **Volcanoes**, such as **Cotopaxi**, occur along the chain of mountains. The **Peru–Chile** trench, which reaches depths of 8000 m, occurs at the point of subduction. **Earthquakes**, such as that in northern Peru in 1970 which killed 67 000 people, are common and are often of high magnitude. There are even **tsunami** threats from shallow earthquakes.

Maps are less useful for supplementing text, but they can be beneficial in identifying the complex physical and human causes of an event, or the spatial impact of a disaster, including deaths and damage.

Always design the map for a specific question. Figure 108 shows a generic map of the Boscastle floods. Selected information has been added to show the causes of this flood event. High-quality annotation with precise facts and figures is very important.

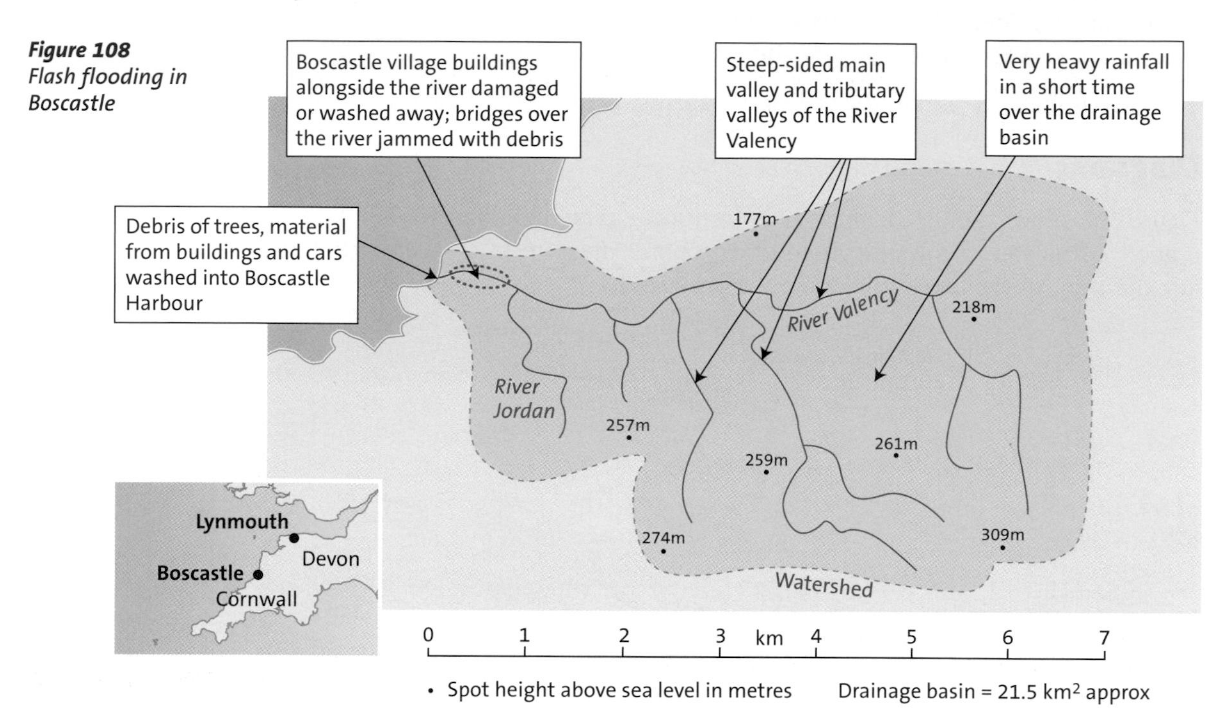

Figure 108
Flash flooding in Boscastle

Drawing maps and diagrams improves your understanding and also that of the reader. Practise drawing key maps and diagrams, to improve your speed.

Planning and writing hazard essays

Deconstructing the title

You have to be able to understand the **command word**. For an A2 essay the following command words are important:

- **Discuss** — investigate, giving evidence for and against the statement
- **Examine** — look closely into the statement, giving detailed support
- **Evaluate/assess** — make a judgement, supported by evidence
- **Explain** — clarify, interpret and account for; give reasons for

Extract the **key words**. These could be:

- **topic** key words (e.g. tectonic hazards)
- **place** and **scale** key words (e.g. LEDC/MEDC or national/local)
- **issue** key words (e.g. factors or role)

A deconstructed title is shown in Figure 109.

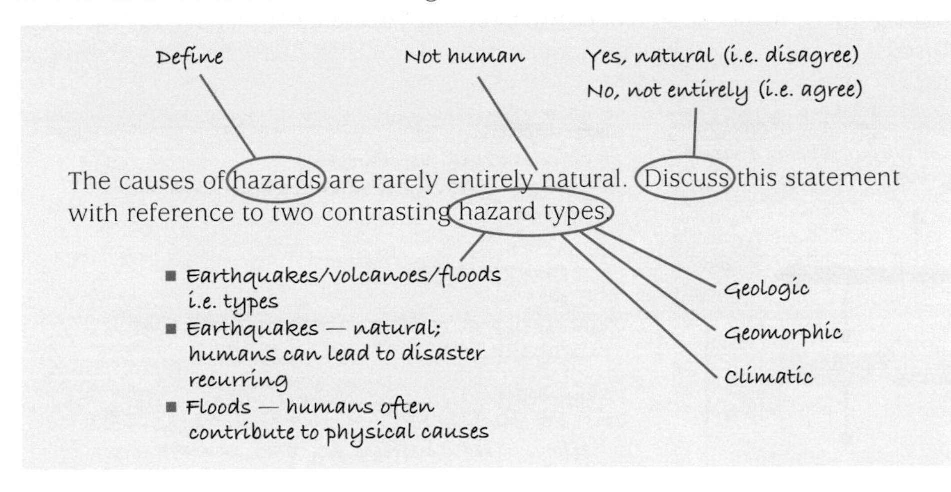

Figure 109
An example of how to deconstruct a title

Planning

After deconstructing the title, you should plan the essay. In the exam, you have about 10 minutes to do this. Most people think around a title using a spider diagram or similar. However, the ideas then have to be classified and sequenced.

Figure 110 is a plan for the deconstructed title given in Figure 109.

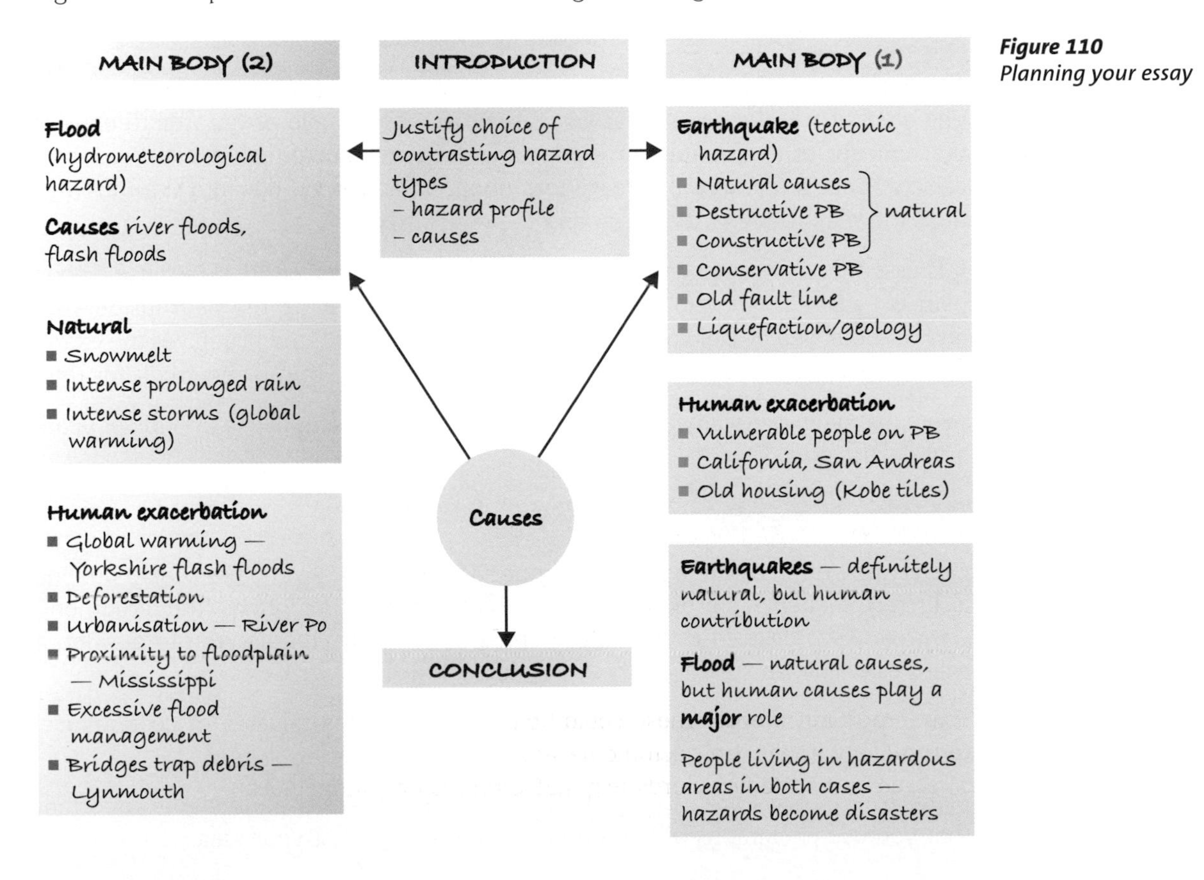

Figure 110
Planning your essay

Figure 111 shows an orderly plan in response to the title: 'The main aim of hazard management should be to reduce the effects of hazards, not manage their causes'. Discuss.

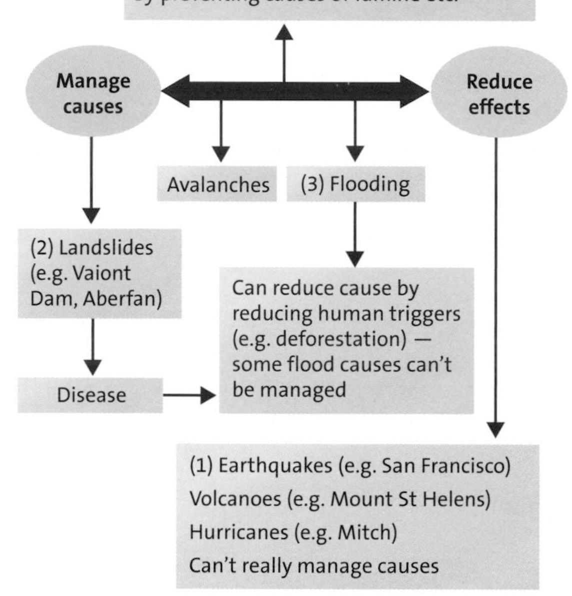

Introduction:

- Define hazard (versus event)
- Causes — management varies according to type
- Effects:
  - social
  - environmental
  - economic

Warning, prep, rescue and recovery to manage vulnerability

Main body:

- Some causes of hazards can't be managed — full efforts to reduce effects, e.g. EQs, San Francisco
- Some causes management could prevent hazard altogether — full efforts to reduce cause, e.g. Vaiont Dam, Aberfan
- Choice of options:
  - better to manage causes, e.g. Bangladeshi floods
  - better to prepare vulnerability management for when causal management fails, e.g. Galtur

Conclusion:

- Better to manage cause where possible — societies have to accept that this can only limit effects however

Figure 111
Planning a hazard essay

The student has devised a neat visual scale across a range of hazards to explore the concept of a spectrum of management from 'manage causes' to 'reducing effects'. This is followed by clear ideas about what can be put into the introduction, the main body of the essay and the conclusion.

It is essential that you practise planning answers to a range of titles, because a good plan is usually the key to a good essay. In the exam, write the plan inside your answer book and do not cross it out. If you run out of time, the plan gives the examiner a clear idea of what your design was and you may gain some credit for this.

Avoid planning case study by case study. This leads to descriptive essays full of death and destruction.

Writing the essay

Quality of written communication is important. Your essay structure should provide a linked sequence of analysis and discussion that maintains the argument. Command words such as 'discuss' and 'to what extent' require you to develop a mature, evaluative style.

It is important to develop good standards of spelling, particularly of geographical terminology and technical words.

Many exam essays are now marked online. This makes it particularly important that you use paragraphs of appropriate length to signpost your ideas.

Index